4-16-14

Evolution and Medicine

Evolution and Medicine

ROBERT L. PERLMAN

OXFORD
UNIVERSITY PRESS

Evolution and Medicine. First Edition. Robert L. Perlman © Robert L. Perlman 2013.
Published 2013 by Oxford University Press.

OXFORD

UNIVERSITY PRESS

Great Clarendon Street, Oxford, OX2 6DP,
United Kingdom

Oxford University Press is a department of the University of Oxford.
It furthers the University's objective of excellence in research, scholarship,
and education by publishing worldwide. Oxford is a registered trade mark of
Oxford University Press in the UK and in certain other countries

British Library Cataloguing in Publication Data

Data available

ISBN 978-0-19-966171-8 (hbk)
ISBN 978-0-19-966172-5 (pbk)

Printed and bound by
CPI Group (UK) Ltd, Croydon, CR0 4YY

To Caryle, who has nurtured me, sustained me,
and enabled me to reach this season

Preface

This book has had its own evolutionary history, with countless generations of revisions and editorial changes. Its final form illustrates the tradeoffs and constraints that are central to the evolutionary process. Time is a major constraint in writing, as it is in life, and choosing how to use my time entailed tradeoffs. Time spent discussing ideas with friends meant time away from writing. And time spent sleeping meant time away from both. The length and scope of the book reflect a tradeoff between being affordable and attractive to readers on the one hand, and being more comprehensive and scholarly on the other. And the book is of course constrained by my own ability to understand and make judgments about difficult and controversial issues in areas in which I have no personal experience or expertise. Readers will have to judge how limiting the constraints have been and how well I navigated the tradeoffs.

One of the most heated controversies in evolutionary biology concerns the origins of altruistic behavior. My only contribution to this debate is to point out that, whatever its origins, altruistic behavior is flourishing in the scientific community. Many friends, colleagues, and people I know only as "e-pen pals" have been wonderfully generous in reading, commenting on, and discussing chapters of this book with me. Among the people who have given me advice, suggestions, and criticism are Julie Bubeck Wardenburg, Don Chambers, Fred Coe, Jeanne Galatzer-Levy, Thomas Grunewald, Yvonne Lange, Ulf Lindahl, Peggy Mason, Mikey McGovern (who also prepared the figures), Rachel Perlman, Ira Salafsky, Alan Schechter, Ben Siegel, Ted Steck, Bernard Strauss, and Bill Wimsatt. They have tried hard to keep me from error. If they have not succeeded, the fault is mine.

The book is a product of the environments in which I have worked. I have been extremely fortunate in being on the faculty at the University of Chicago. The U of C, as it is known locally, is an extraordinarily rich and intellectually active community. It provides the faculty with outstanding students and colleagues, unbelievable library and research support, and, most importantly, the freedom to pursue their own ideas and interests. I couldn't have asked for a more stimulating environment. Several years ago, I had the good fortune to be a fellow at another extraordinary institution, the Wissenschaftskolleg in Berlin. I am especially grateful to Randy Nesse and Joachim Nettelbeck for inviting me to the Wiko.

Chance, too, played an important role in the evolution of this book. Sometime after I had begun writing but still wasn't sure how to find a publisher, I happened to be at a meeting with Gillian Bentley and happened to mention that I was writing a book on evolutionary medicine. Gillian put me in touch with Helen Eaton, who was then at Oxford University Press, and through this contact the book came into its present form. Helen and Ian Sherman were enthusiastic and helpful from the start, and Lucy Nash picked up where Helen left off, keeping me on track and cheering me up when I was feeling bogged down. Lucy, too, has become an e-pen pal. I am of course delighted that my book is being published by Oxford. It is sobering to realize that without Gillian's timely intervention, the book might well have had a different fate.

Even though OUP was interested in my book, I needed to negotiate a publishing agreement with them. Ivan Dee and Arthur Mason taught me how to read and understand these agreements. Without their help, I might still be worrying about movie rights.

I have a special debt to my family and an extra special debt to my wife, Caryle. The time I spent writing the book was time away from her. Caryle was unfailingly supportive of my writing even though it meant that I was often preoccupied and distracted, and it caused us to forgo a number of activities we would have enjoyed. Her love, patience, and encouragement are just a few of the many reasons I have dedicated this book to her. Finally, my children, Rachel and Ezra, their partners, Tamara and Shireen, and their children, Zach, Sam, Maelis, and Marella, are constant reminders that we have a future worth preserving, and that a book that might help improve our future was worth writing. Thank you, all.

Contents

Abbreviations   xii

1. Evolution and medicine   1

   1.1   Introduction   1

   1.2   The theory of evolution by natural selection   3

   1.3   The different conceptual bases of medicine and evolutionary biology   7

   1.4   Why our evolutionary heritage has left us vulnerable to disease   11

2. Human demography, history, and disease   13

   2.1   Introduction   13

   2.2   Population growth: birth rates and death rates   13

   2.3   Population growth in age-structured populations: fertility rates   15

   2.4   Age-specific death rates   19

   2.5   History of human population growth   21

   2.6   The future of the human population   26

3. Evolutionary genetics   29

   3.1   Introduction   29

   3.2   Other evolutionary processes: mutation, genetic drift, and migration   30

   3.3   Genetic dominance   33

   3.4   Heterozygote advantage   34

   3.5   Pleiotropy and epistasis   35

   3.6   Linkage and hitchhiking   36

   3.7   Frequency dependent selection   36

   3.8   Epigenetic regulation of gene expression   37

   3.9   Population structure and mating patterns   38

   3.10 Genetic consequences of human evolutionary history   39

   3.11 Natural selection in human populations   40

4. Cystic fibrosis   43

   4.1   Introduction   43

   4.2   CFTR, the cystic fibrosis transmembrane conductance regulator   45

   4.3   Genotypic diversity and phenotypic heterogeneity in cystic fibrosis   46

   4.4   Relationship between genotype and phenotype   47

   4.5   Evolution of mutant CFTR alleles   48

5. Life history tradeoffs and the evolutionary biology of aging   51

   5.1   Introduction   51

   5.2   The causes of death change through the life cycle   53

   5.3   What is aging?   54 x Contents

5.4	The life history theory of aging	55
5.5	Genetic causes of aging	57
5.6	Proximate causes of aging	58
5.7	Somatic repair and the depletion of physiological capital	59
5.8	Plasticity in rates of aging	61
5.9	Developmental origins of health and disease	62

6. Cancer — **65**

6.1	Introduction	65
6.2	Cancer as a disease of aging	65
6.3	Regulation of cell growth and replication	66
6.4	Selection for cells that escape normal growth controls	69
6.5	Cancer progression	71
6.6	Ecology of cancers	72
6.7	Anti-cancer defenses	73
6.8	Carcinogenesis and cancer prevention	74

7. Host–pathogen coevolution — **77**

7.1	Introduction	77
7.2	Epidemiology of pathogen transmission	78
7.3	Virulence and transmissibility	80
7.4	Host–pathogen coevolution: hosts evolve in ways that minimize the fitness cost of pathogens	81
7.5	Host–pathogen coevolution: pathogens evolve in ways that optimize their fitness	82
7.6	Myxomatosis: a case study of host–pathogen coevolution	84
7.7	Complexities in host–pathogen interactions	85
7.8	Antibiotic resistance: methicillin–resistant *Staphylococcus aureus*	87
7.9	Manifestations of disease	90

8. Sexually transmitted diseases — **91**

8.1	Introduction	91
8.2	The epidemiology of sexually transmitted diseases	92
8.3	Evolutionary responses of hosts to sexually transmitted pathogens	94
8.4	Syphilis	95
8.5	HIV/AIDS	98

9. Malaria — **103**

9.1	Introduction	103
9.2	The life history of *Plasmodium falciparum*	103
9.3	The natural history of malaria infections	106
9.4	R_o, the basic reproductive number of *P. falciparum*	108
9.5	Virulence of *P. falciparum*	109
9.6	Evolution of *P. falciparum* and other malaria parasites	110
9.7	Mosquitoes that transmit *P. falciparum*	111
9.8	Effects of malaria on human evolution	112
9.9	The future of malaria	114

10. Gene–culture coevolution: lactase persistence   115

    10.1 Introduction   115

    10.2 Milk consumption   116

    10.3 A brief history of animal domestication and dairying   117

    10.4 The evolution of lactose synthesis and metabolism   118

    10.5 Lactase restriction and lactase persistence   120

    10.6 The coevolution of lactase persistence and dairying   122

11. Man-made diseases   127

    11.1 Introduction   127

    11.2 Diet, obesity, and diabetes   128

    11.3 Salt intake and hypertension   131

    11.4 Elimination of old pathogens: the hygiene hypothesis   134

    11.5 Hierarchical societies and socioeconomic disparities in health 137

    11.6 Reducing the burden of man-made diseases   141

Glossary   143

References   145

Index   159

Abbreviations

CDK	cyclin-dependent kinase
CFTR	cystic fibrosis transmembrane conductance regulator
CMAH	CMP-Neu5Ac hydroxylase
EEA	environment of evolutionary adaptedness
ENCODE	Encyclopedia of DNA elements
HbS	sickle cell hemoglobin
HLA	human leukocyte antigen
HPV	human papilloma virus
MHC	major histocompatibility complex
MRSA	methicillin-resistant *Staphylococcus aureus*
Neu5Ac	N-acetyl-neuraminic acid
Neu5Gc	N-glycolyl-neuraminic acid
OMIM	Online Mendelian Inheritance in Man
PBP2	penicillin binding protein 2
PfEMP1	*P. falciparum* erythrocyte membrane protein 1
pRb	retinoblastoma protein
SIV	simian immunodeficiency virus
STD	sexually transmitted disease
TFR	total fertility rate
TGFβ1	transforming growth factor β1
TprK	*T. pallidum* repeat protein K
UNAIDS	U.N. Program on HIV/AIDS

Medical education does not exist to provide students with a way of making a living, but to ensure the health of the community.

RUDOLF VIRCHOW

An evolutionary view of human health and disease is not surprising or new; it is merely inevitable in the face of evidence and time.

CHARLES R. SCRIVER (Scriver 1984)

1

Evolution and medicine

1.1 Introduction

Charles Darwin "had medicine in his blood" (Bynum 1983). His father and grandfather were physicians, and he himself studied medicine. Although Darwin left medical school after two years and did not become a physician, he retained a strong interest in medicine and regularly used examples drawn from human biology and medicine in his writings. Clearly, he believed that medicine fell within the purview of his theory of evolution and he recognized the ways in which the study of evolution and of medicine could be mutually enriching. In *The Descent of Man*, Darwin argued that humans, like other species, have evolved from earlier, ancestral species (Darwin 1871). "Descent with modification," Darwin's term for evolution, accounts for the many anatomic and physiologic similarities between humans and other animals. Rudimentary organs played an important role in Darwin's argument. These organs have no function in humans and, as with the appendix, they may increase the risk of disease and death. They can only be understood as relics of structures that had a function in our evolutionary ancestors and that have decreased in size but have not been eliminated during human evolution. Darwin was especially interested in heritable variation, which plays a central role in his theory of evolution by natural selection. In *The Variation of Animals and Plants under Domestication* (1883), Darwin discussed heritable variation in humans. After mentioning a number of trivial or unimportant variations, such as families in which several members had one lock of hair that was differently colored from the rest, he noted that there are also inherited variations in predispositions to various diseases and he then discussed heritable diseases of the eye in detail (Darwin 1883, vol. 1, pp. 452–54).

As the theory of evolution became more widely known and accepted in the late nineteenth century, some physicians began to apply evolutionary concepts to medicine (Bynum 1983; Zampieri 2009). For the most part, however, these efforts had little lasting impact. Perhaps the most important contribution of evolutionary thinking to medicine in the nineteenth century was the work of the neurologist John Hughlings Jackson (Jackson 1884). Jackson viewed both the development of the nervous system and the loss of function in neurological diseases from an evolutionary perspective. He saw the evolution of the nervous system as progressive, beginning with the automatic or involuntary regulation of respiration and circulation, and culminating in the "highest centres" of consciousness and mind, which controlled the lower centers. Jackson noted that these highest, and evolutionarily most recent, portions of the brain were most susceptible to damage by neurotoxins (alcohol, for example) or disease

Evolution and Medicine. First Edition. Robert L. Perlman © Robert L. Perlman 2013.
Published 2013 by Oxford University Press.

(epilepsy), and thus many neurological diseases resulted in what he called "dissolutions," or reversals of evolution. Jackson's views on the hierarchical, evolutionary organization of the nervous system continue to influence thinking in neurology. For example, Paul MacLean's concept of the triune brain proposes that the human brain comprises a reptilian brainstem, an early mammalian limbic system, and a more recent neocortex (MacLean 1990). But Jackson's ideas have had relatively limited impact on other branches of medicine.

In the mid-twentieth century, the British biologist J. B. S. Haldane suggested that "the struggle against disease, and particularly infectious disease, has been a very important evolutionary agent" (Haldane 1949a). Haldane and Anthony Allison, a physician interested in parasitology and tropical medicine, independently proposed what became known as the malaria hypothesis. Specifically, they proposed that the alleles that cause the diseases thalassemia and sickle cell anemia spread in human populations because, when these alleles were present in heterozygous individuals, they conferred resistance to malaria (Allison 1954; Haldane 1949b). Allison went on to demonstrate that people who were heterozygous for sickle cell hemoglobin were in fact resistant to malaria, and that the selective advantage of malaria resistance could account for the frequency and geographic distribution of the sickle cell trait (Allison 1964). Although Haldane's insight and Allison's research stimulated a search for other genetic variants that were maintained because they conferred resistance to malaria, such as glucose-6-phosphate dehydrogenase deficiency (Luzzatto et al. 1969), they did not lead to a broader incorporation of evolutionary thinking into medicine.

The emergence of antibiotic-resistant bacteria shortly after the introduction of antibiotics into clinical medicine is the most striking example of the medical relevance of evolution (Dubos 1942). Concerns about antibiotic resistance led to important studies on the mechanisms of resistance and to the development of new antibiotics that overcame this resistance. Recognition that the spread of antibiotic-resistant bacteria was due to selection for antibiotic resistance led to calls for the more responsible use of these drugs. Unfortunately, these calls largely went unheeded. Moreover, little attention was given to understanding the dynamics of selection or the ways in which regimens of antibiotic usage might modulate the strength of selection for antibiotic resistance.

Until recently, the hierarchical organization of the nervous system, the prevalence of disease-associated alleles, and the spread of antibiotic resistance were simply isolated instances of the application of evolutionary concepts to medicine. Stimulated by the pioneering publications of Randolph Nesse and George Williams in the 1990s, however, physicians and other scientists have now begun to integrate evolutionary biology and medicine into a coherent discipline (Nesse and Williams 1994; Williams and Nesse 1991). This is the new field of Darwinian, or evolutionary, medicine (Gluckman, Beedle, and Hanson 2009; Stearns and Koella 2008; Trevathan, Smith, and McKenna 2008).

Given that the theory of evolution by natural selection is the central, unifying theory in biology and that our understanding of disease is heavily based on our knowledge of human biology, it may seem surprising that evolutionary medicine is such a new field. There are many reasons why evolutionary biology and medicine developed as separate disciplines and have until recently remained isolated from one another. When Darwin proposed his theory of evolution by natural selection, medicine was already a well-established profession, with a history in the West going back at least 2500 years to Hippocrates. In the nineteenth century, medical practice stressed careful physical examination of patients, description of the natural histories of diseases, and correlation of the signs and symptoms of disease with autopsy findings. Later, with the rise of the germ theory of disease, medicine became increasingly focused on laboratory diagnoses and on identifying the etiologies or causes of disease (Porter 1998). Medicine was taught in its own institutions, which were typically based in hospitals, and the medical curriculum was already crowded. There was no room and no apparent need to bring the theory of evolution into medical education, research, or practice.

Evolutionary biology did not develop into an academic discipline until long after Darwin. At the time of the Flexner Report (Flexner 1910), which laid the foundations for today's science-based medical education, there were still no university departments, professional societies, or scholarly journals devoted to evolution. Only after the integration of evolutionary biology with genetics in the 1930s and 1940s did evolutionary biology become a mature science (Ruse 2009). Even then, evolutionary biology and medicine continued to develop as separate disciplines, with little interaction. Evolutionary biologists were concerned with classification of species, with enriching and analyzing the fossil record, and with finding evidence of natural selection in the wild. Except for paleontological studies of human origins, most evolutionists shied away from human biology. Many of these biologists worked in museums and field stations, isolated from medical centers, and they may not have wanted to be associated with the eugenics programs of the early twentieth century that had been embraced by some evolutionists (Kevles 1995). Perhaps most importantly, as the following brief review of the theory of evolution by natural selection will make clear, evolutionary biology and medicine have different and seemingly incompatible ways of understanding biological phenomena. Evolutionary biologists and physicians have been concerned with different problems, they speak different specialized languages, and they see the natural world in different ways. These differences have helped to keep these fields apart and continue to hinder their integration.

1.2 The theory of evolution by natural selection

Although our understanding of evolution has increased greatly since Darwin's time, biologists still use essentially the same argument to support the theory of evolution by natural selection as Darwin did when he proposed it.

Darwin began by pointing out the abundant variation that exists among individual organisms in a population. The first two chapters of *On the Origin of Species* are devoted to a discussion of variation, first in domesticated species and then in nature (Darwin 1859). Darwin focused on small, often barely discernible, variations; he regarded the greatly deviant organisms that occasionally arise in nature as "monstrosities" that had no role in evolution. Of course, people had long been aware of variations among organisms within populations or species. As Ernst Mayr (1964) has emphasized, however, before Darwin species were understood in typological or essentialist terms. In this view, each species was thought to be characterized by a unique, unchanging essence. Variation was seen as an irrelevant distraction, due to imperfections in the material realization of the ideal form of the species (Mayr 1964). Darwin introduced what Mayr called "population thinking" into biology. Biologists no longer think of species as having ideal or essential forms. Instead, they commonly think about species (at least extant, sexually reproducing species) in terms of Mayr's biological species concept. According to this concept, species comprise populations of organisms that can interbreed and produce viable offspring in nature but that otherwise exhibit a wealth of variation and change over time—in other words, species evolve (Mayr 1988b). Variation remains a critical aspect of evolutionary thinking because it provides the raw material for evolution by natural selection.

Next, Darwin pointed out that, while the number of organisms in a population might potentially increase without limit, the resources needed to support these populations are finite. In other words, the reproductive capacity of the organisms in a population must greatly exceed what we now call the carrying capacity of the environment, the population that the local habitat can sustain. This inequality between reproductive potential and environmental resources means that individual organisms in a population must compete for survival and reproduction. Darwin called this competition the "struggle for existence." Darwin based this concept on Thomas Malthus's *Essay on the Principle of Population* (Malthus 1798); he refers to the struggle for existence as "the doctrine of Malthus applied with manifold force to the whole animal and vegetable kingdoms" (Darwin 1859, p. 63). Malthus was concerned with the disparity between human population growth and the availability of food. Darwin expanded Malthus's ideas from humans to all species and from food to all of the environmental resources that organisms need to survive and reproduce. Evolutionists understand the struggle for existence in what Darwin called "a large and metaphorical sense" (p. 62). It refers to all of the difficulties that organisms must overcome in order to survive and reproduce in the complex and challenging environments in which they live. Organisms struggle to secure food and other resources they need to grow and develop, to avoid being eaten by predators, to attract mating partners and reproduce, and to promote the survival of their offspring. The struggle for existence is primarily a struggle between organisms and their environments. The term may conjure visions of hand-to-hand combat but only occasionally does the struggle for existence involve a direct physical confrontation between two individuals of the same species, as in two dogs fighting over a scarce piece of meat or two males fighting to mate with a female.

The environment in which the struggle for existence takes place includes both the physical or nonliving environment (air, water, sunlight, climate, etc.) and the living or biotic environment. The biotic environment comprises all of the other species with which organisms interact or on which they depend (directly or indirectly), as well as other members of their own species. Organisms of other species constitute especially important components of an organism's environment. For this reason, evolution is closely connected to ecology and to the ecological relationships among species. Many of us in developed countries live in environments in which our interactions with organisms of other species are largely hidden. Our direct experience is limited to our pets, to the plants and animals in our gardens and parks, to the insects and other pests that annoy or plague us, to infectious microorganisms, and to the foods we eat, many of which we purchase pre-packaged in grocery stores. We should remember, however, that our lives and our health are intimately related to and affected by the innumerable species that form part of our environment—those that contribute to our health as well as those that cause disease.

Those individuals that are successful in the struggle for existence will survive, reproduce, and leave offspring; in evolutionary terms, producing offspring who themselves survive and reproduce is the definition of success. Biologists commonly use the term fitness, sometimes modified as reproductive or evolutionary fitness to avoid confusion, to denote this reproductive success. The term "survival of the fittest," introduced by the English philosopher Herbert Spencer, has become a widely used metaphor to describe the evolutionary process (Spencer 1864). This metaphor may be misleading, however, because it is easy for people who are concerned with "fitness" today to think that evolutionary fitness refers to something akin to physical fitness. In evolutionary terms, fitness does not simply refer to strength or endurance, but to all of the traits that enable organisms to function—to survive and produce offspring—in their environments. A more appropriate meaning of fitness might be suitability. Successful organisms are well suited to their environments. They fit into and may shape their environments the way hands fit into and shape gloves.

"Survival of the fittest" may also be misleading because it seems to imply that fitness is an attribute of individual organisms. Although we often talk loosely about the fitness of organisms, fitness is best understood in terms of alleles or genotypes. Fitness is the expected average reproductive success of organisms of a given genotype, relative to the average reproductive success of other organisms in the population. Alleles that enhance fitness survive in the sense that they are preferentially transmitted from parents to offspring. In genetic terms, fitness may be thought of as the ability of organisms of a specific genotype to contribute genes to the gene pools of their populations. Organisms can pass on their genes directly, by their own reproduction, or indirectly, by enhancing the reproductive success of their genetic relatives. A broader concept of fitness, which is especially relevant to social species such as humans, is inclusive fitness, which comprises both the direct and indirect components of fitness (Hamilton 1964).

Although Darwin did not understand the molecular basis of heredity, he recognized that many traits are heritable. By and large, offspring tend to resemble their parents. As a result, traits that increase survival and reproduction—traits that make organisms well suited to their environments and thus enable them to succeed in the struggle for existence—will in general spread in the population. In contrast, traits that decrease survival and reproduction, and the alleles that underlie these traits, will, over time, be eliminated. This is natural selection, which Darwin defined as "This preservation of favourable variations and the rejection of injurious variations" (Darwin 1859, p. 81). Favorable variations—traits associated with increased fitness—that are preserved by natural selection are known as adaptations. Darwin adopted the term natural selection by analogy with artificial selection, which he called "selection by man." Natural selection may also be a misleading term, since it implies that nature, like humans, is actively selecting the traits that spread in populations. It may be more appropriately understood as a process of nonrandom elimination of organisms, along with their traits and their genes. Darwin rarely used the word "evolution," which originally meant unrolling or unfolding. In the nineteenth century, evolution was commonly used to describe development, which was thought to result from the unfolding of a pre-existing developmental plan. Instead, as mentioned earlier, Darwin referred to evolution as "descent with modification."

All that is needed for Darwinian evolution, or evolution by selection, is a population of entities that exhibit heritable variation in traits that affect their reproductive success, their success in leaving progeny who themselves survive and reproduce. Since populations of living organisms have these properties, evolution by natural selection is inevitable (Lewontin 1970). Other entities that have these properties, including computer viruses, cultural traits, and artificial organisms, may evolve by selective mechanisms that are analogous to natural selection. Artificial selection, or selection by humans, continues to shape the evolution of domesticated species of plants and animals, as well as the evolution of antibiotic resistance in bacteria and other pathogens. Natural selection may be thought of as a natural law of biology; it is a necessary consequence of the nature of living organisms.

Evolution, however, is a historical process, which depends on chance events and historical contingencies as well as on natural selection. For this reason, the course of evolution is not predictive in the way that some physical laws are. As the French biologist Jacques Monod has written, biological processes result from "chance and necessity" (Monod 1971). Natural selection plays a special role in evolution because it is the process that gives rise to adaptations, to traits that enhance reproductive fitness. Despite the attention that is understandably given to natural selection, however, we should not forget or minimize the importance of chance in evolution.

An important component of natural selection is sexual selection, which results from competition among members of the same sex for access to mating partners, often through being chosen by members of the opposite sex (Cronin 1991). The peacock's tail is the classic example of a trait that arose and is maintained by sexual selection. Large, brightly colored tails attract predators and decrease the survival of peacocks. These large tails evolved because peahens preferred to mate with peacocks that had them, thereby increasing the reproductive success of these peacocks. Many human traits, including patterns of death and disability, are thought to have evolved as a result of sexual selection (Kruger and Nesse 2004).

Evolution by natural selection begins with the presence of heritable variations among individual organisms. Organisms that have favorable variations will (relative to organisms without these variations) survive, reproduce, and transmit these traits to their offspring, and so adaptations, traits that increase reproductive success, will spread in a population. Equally importantly, traits that reduce reproductive success—Darwin's "injurious variations"—will decrease in frequency. For the most part, evolution involves the gradual accumulation and summation of many small variations. As a result, the production of adaptations is a slow process, typically taking many, many generations. If two populations of a species evolve in different environments, they will slowly come to differ, both because different traits will enhance fitness and be selected in different environments, and because of chance events that occur in one population but not the other. As these populations diverge to the point that they are recognizably different, they will generally be referred to as different varieties or subspecies. And as they diverge further, organisms from the two populations may no longer mate with one another because of physical, biochemical, or behavioral differences—or, if they do mate, they may not produce viable and fertile offspring. At this point, biologists would say they have evolved into different species. Biologists frequently distinguish between microevolution, evolutionary changes within a species that lead to the spread of adaptations and the production of distinct varieties or subspecies, and macroevolution, the formation of new species or higher taxa. As Darwin argued, when microevolutionary processes are continued over long time periods, they can eventually lead to macroevolution. Adaptations to different environments often underlie the origin of species.

1.3 The different conceptual bases of medicine and evolutionary biology

Medicine and evolutionary biology bring markedly different perspectives to the study of biological phenomena. Medicine has traditionally focused on individuals. Physicians are concerned with the health and well-being of their individual patients. Their primary goal is to keep their patients healthy. When their patients do get sick, physicians are interested in diagnosing their patients' diseases and in understanding how these diseases cause the symptoms that they do, because they wish to restore their patients to health or at least relieve their discomfort. Only in times of epidemics are physicians concerned with the spread of disease in populations and with ways in which they might help their patients avoid these diseases. In contrast, evolutionary biology focuses on populations or species. Evolutionists are interested in variations within populations and the ways in which populations change over time. Individual survival and reproduction are crucial for evolution. Differences in the survival and fertility of individuals—differences in fitness of organisms with different genotypes—provide the basis for evolutionary change. But individuals are born, develop, progress through a life cycle, and die. Only populations evolve.

Physicians and evolutionists use different metaphors to describe and understand their work. One of the most common metaphors for medicine is war; we talk about diseases as enemies and our therapeutic armamentarium as weapons. Richard Nixon's "war on cancer" is just one of the wars we have declared against disease. Sometimes we are unaware of these metaphors; as the British physician Paul Hodgkin pointed out, a "cohort," which is now used to describe a group of subjects in a clinical trial, was originally a group of soldiers in a Roman legion (Hodgkin 1985). The popularity of the "medicine is war" metaphor is not surprising, since modern therapeutics developed in the shadow of World War II and the Cold War. But the uncritical adoption of this metaphor, with patients as the battleground rather than the focus of medical attention, may lead physicians to carry out actions that are not in the best interests of their patients (Hodgkin 1985).

Evolutionary biologists also use military metaphors. Host–pathogen coevolution is often described as an "evolutionary arms race." But evolutionary concepts are more commonly expressed in economic than in military terms; the parallels between ecology and economy run deeper than etymology. Karl Marx, who had a high regard for Darwin and his work, was perhaps the first person to realize this. As he commented, "It is remarkable how Darwin rediscovers, among the beasts and plants, the society of England with its division of labor, competition, opening up of new markets, 'inventions' and Malthusian 'struggle for existence'" (Marx 1862). It is also not surprising that Darwin was influenced by Adam Smith and other British economists, and by the intellectual climate of Victorian England (Lewontin 1990; Schweber 1980). Metaphors such as struggle for existence and survival of the fittest are essential in helping us understand abstract concepts (Lakoff and Johnson 2003). But a failure to appreciate the ways in which metaphors shape our thinking can be problematic. We have already discussed some of the confusions caused by the metaphors of struggle and fitness. And as several authors have pointed out, the focus on competition in evolutionary thinking has hindered acceptance of the roles of cooperation and symbiosis in evolution (Ryan 2001; Sapp 1994; Weiss and Buchanan 2009).

Because of their concern for their individual patients, physicians develop expertise at synthesizing and integrating their patients' medical, personal, and family histories, their symptoms, the findings of physical examinations, and the results of laboratory tests. This deep understanding of patients, and the relationships that develop in the process of gaining this understanding, is an integral part of medical care. The diagnostic process in medicine is similar to the process of arriving at evolutionary explanations. Both require judgments about the ways that historical events have resulted in present conditions and both depend on abduction, or reasoning to the most likely explanation. But medical therapeutics is guided by controlled trials of a kind that are seldom possible in evolutionary biology. Evolutionists are concerned about changes in populations over time and their research typically requires the creation of quantitative mathematical models to test hypotheses about the mechanisms and rates of these changes. The standards of evidence that are relevant to evolutionary experiments are totally different from those of evidence-based medicine. The different subject matters of medicine and evolutionary biology lead their practitioners to develop different intellectual styles.

The most widely cited definition of health, as developed and promulgated by the World Health Organization (WHO), is not merely the absence of disease or infirmity but "a state of complete physical, mental, and social well-being" (World Health Organization 2006). More recent definitions have stressed the abilities of individuals to adapt and self manage in the face of social, physical, and emotional challenges (Huber et al. 2011). Natural selection, however, acts to maximize the reproductive success of organisms, not their well-being or their ability to self manage in the face of challenges. Selection may result in longevity and health but these outcomes are byproducts of selection for increased reproductive fitness. Organisms have to live long enough and be healthy enough to reproduce and to promote the survival of their offspring, but that is all. Evolutionary fitness is not the same as health.

Physicians and their patients confront tradeoffs and constraints regularly, when they are forced to weigh the risks, benefits, and costs of treatment options, but they usually view these tradeoffs as practical problems rather than as inescapable facts of life. The notion of health as a "state of complete . . . well-being" does not carry any acknowledgment that tradeoffs may prevent the attainment of this goal. In contrast, evolutionists recognize that tradeoffs and constraints limit the ability of natural selection to optimize fitness and believe that they play a large role in evolutionary processes.

Individual organisms are the products of two distinct histories—their own life history, or ontogeny, and the evolutionary history of their species, or phylogeny. Biologists often divide the causes of biological phenomena into proximate causes, causes that operate during the lifetime of an individual, and ultimate causes, causes that operated during the evolutionary history of the species (Mayr 1988a). Proximate causes are sometimes said to answer "how" questions—for example, how (by what physiological mechanisms) do we raise our body temperature in response to infection—while ultimate causes answer "why" questions—why (for what evolutionary reasons) do we have a febrile response to infection? The Dutch ethologist Nikolaas Tinbergen (1963) pointed out that traits have two distinct proximate causes and two ultimate causes. The proximate causes of a trait include its development during an organism's ontogeny and the physiological or molecular mechanisms that produce it; the ultimate causes are its phylogenetic origin and its adaptive significance (Tinbergen 1963). Physicians have traditionally been concerned with proximate causes of disease because these are the causal pathways that are amenable to medical intervention. In contrast, evolutionists want to understand ultimate causes of biological phenomena. Recent advances in evolutionary development biology, or evo-devo, have called attention to the relationship between evolution and development, and have led to a blurring of the distinction between proximate and ultimate causes (Laland et al. 2011). As we shall discuss later, there is currently great interest in understanding the ways in which our evolved mechanisms of development may predispose us to disease in adult life.

Physicians focus on the health of human beings. To a great extent, medicine has tried to separate humans from the rest of nature and protect us from species that might cause disease. Evolutionists, on the other hand, view populations as embedded in ecological communities that comprise a myriad of interrelated and interacting species, all of which are subject to natural selection and are therefore coevolving. Physicians certainly recognize environmental causes of disease, especially infectious diseases and diseases due to environmental toxins. Nonetheless, medical research has focused on the inner workings of human beings, on the physiological and pathophysiological mechanisms that promote health or lead to disease. Medicine is concerned with what Claude Bernard termed the internal environment, the blood and extracellular fluids that provide the immediate environment in which our cells and organs function (Bernard 1957). In this view, health involves the maintenance of constant, or nearly constant, conditions in the internal environment—conditions that enable cells and organs to function properly—while diseases are manifest by deviations from these "normal" conditions. Evolutionary biologists appreciate that the physiological mechanisms that maintain homeostasis are adaptations that enhance fitness but they are more interested in studying the interactions of organisms with their external environments, because it is these ecological interactions that shape the struggle for existence and natural selection. Appreciation of the physiological functions and pathophysiological effects of the human microbiome, the communities of microorganisms that inhabit our skin, intestines, and other body cavities, has led to the recognition that humans are ecological communities. Study of the microbiome is another growing area of research in which the interests of physicians and evolutionists are converging (Turnbaugh et al. 2007).

Finally, medicine and evolutionary biology have different ways of thinking about variation. Medicine focuses on notions of normality and abnormality. Physicians distinguish between "normal" values of traits, values that are associated with good health or that are common in the population, and "abnormal" values, values that are associated with an increased risk of disease. In a medical context, this distinction between normal and abnormal often makes good sense. Many deviations from normal values—elevated blood pressure, blood cholesterol, and body mass index, for example—are risk factors for diseases that may be prevented or postponed by medical interventions. Occasionally, however, extreme values of a trait— short stature, for example—may be labeled abnormal even if they do not have implications for health. Since the rise of the Human Genome Project, physicians are certainly aware of and concerned about genetic variations among their patients. But medicine is still influenced by an essentialist view of biology that tends to view phenotypic variations as deviations from a normal, healthy, or ideal state. This medical understanding of variation differs from that of evolutionary biologists, who view variation as a fundamental property of biological populations. Not only is variation abundant in nature, it provides the substrate for evolution by natural selection; if there weren't heritable variations among individuals, populations couldn't evolve. The values of specific traits among individuals typically exhibit a distribution, frequently a normal or lognormal distribution, that is associated with variations in fitness. Often, but not always, the median or mean value of a trait is maintained by natural selection because it is associated with maximal fitness. Only rarely if ever are there sharp cutoffs that separate health from disease or distinguish different levels of fitness.

Historically, then, medicine and evolutionary biology have been concerned with different biological problems and have developed different approaches to study their areas of interest. It is not surprising that they developed as separate, unrelated disciplines. Only recently have physicians and nonmedical biologists begun to realize that there is much to be gained by integrating these disciplines. Evolutionary medicine is based on the recognition that these different perspectives are not mutually exclusive but complementary, and that integrating them will give a richer understanding of health and disease. Understanding evolutionary processes helps to explain our evolved vulnerabilities or susceptibilities to disease and our current burden of disease. Conversely, since disease has served as an important selection factor in evolution (Haldane 1949a), knowledge of the present patterns of disease gives insights into our evolutionary history. Analysis of the evolutionary causes of diseases may lead to novel strategies to prevent, postpone, or ameliorate them. Understanding both the proximate and ultimate causes of diseases will provide a richer understanding of disease. Finally, evolutionary explanations of disease are important because patients often want to know why they have the diseases they have. In the absence of evolutionary explanations, they may fall back on unhelpful folk beliefs, such as the idea that their diseases are punishment for sinful behavior (Bynum 2008, p. 18).

1.4 Why our evolutionary heritage has left us vulnerable to disease

Many diseases cause premature death (death before the end of the reproductive and child-raising periods) or reduced fertility. But most diseases do not affect all members of a population or do not affect everyone to the same degree. Rather, individuals exhibit variation in resistance or response to diseases, just as they exhibit variation in virtually all other traits. At least some of this variation is due to genetic or heritable variation in the population. Heritable variations in resistance to these diseases represent variations in fitness; individuals who survive and remain fertile in the face of a disease will on average produce and raise more children than will people who die from or become infertile as a result of the disease. As a disease spreads through a population, natural selection will increase the frequency of alleles that are associated with resistance to it. The alleles associated with resistance to malaria are classic examples of this process.

Despite selection for disease resistance throughout our evolutionary history, however, natural selection has clearly not eliminated disease. Evolutionary medicine helps us understand the limits as well as the power of natural selection in shaping human biology and the reasons—the ultimate causes—for our continued vulnerability or susceptibility to disease. Broadly speaking, there are several important limits to natural selection that contribute to the persistence of disease (Nesse 2005; Perlman 2005).

First, there are limitations intrinsic to the process of evolution by natural selection itself. Diseases that cause premature death or reduced fertility will select for and increase the frequency of alleles that are associated with disease resistance. But natural selection is not the only mechanism of evolutionary change. New alleles can enter populations either by mutation or by gene flow from other populations of the same species. Once these alleles enter a population, their fate is determined by genetic drift (changes in allele frequency due to random sampling in the transmission of alleles from one generation to the next) as well as by natural selection. These other evolutionary processes may counteract the effects of selection by introducing or increasing the frequency of alleles associated with susceptibility to disease. For these and other genetic reasons, beneficial alleles—specifically, alleles associated with disease resistance or a decreased risk of disease—may not spread or become fixed in a population.

Natural selection increases the frequency of traits that enhance reproductive fitness. As we have discussed, however, fitness is not the same as health or longevity. If diseases do not decrease reproductive success, there will not be selection for resistance to them. Diseases of aging, diseases that increase in prevalence after the end of our reproductive and child-raising years, are one class of diseases that may not significantly decrease fitness. Evolutionary life history theory and the evolutionary theory of aging provide a framework for understanding and, possibly, for postponing these diseases.

Natural selection is a slow process. Even when selection is intense, allele frequencies in populations change only gradually over many generations. Change in a population's environment is typically more rapid than genetic change. The other species with which we interact, and especially the pathogens or parasites that infect us and cause disease, constitute an important and rapidly changing component of our environment. Our pathogens, too, evolve by natural selection. Just as our evolutionary ancestors evolved and we are continuing to evolve increased resistance to our pathogens, these pathogens have evolved, and are evolving, to overcome this resistance and to grow in and be transmitted among us. This process of host–pathogen coevolution helps to rationalize the natural histories of infectious diseases and to explain why some infections are relatively benign while others are virulent. Understanding pathogen evolution and host–pathogen coevolution may suggest strategies for slowing the spread of antibiotic resistance and for reducing the virulence of infectious diseases.

The human environment is strongly influenced by cultural beliefs, practices, and artifacts, all of which are subject to rapid change. Disease may result from an inability of natural selection to keep pace with a changing cultural environment—in other words, from a mismatch between the environment in which we now live and the genes we have inherited from our evolutionary ancestors, genes that enabled these ancestors to survive and reproduce in the various environments in which they lived. The increasing prevalence of obesity and hypertension exemplifies the principle that genes which enhanced the fitness of our ancestors may now increase our risk of disease.

There are several other constraints on natural selection. In brief, macroevolution constrains microevolution (Stearns et al. 2008). Our macroevolutionary history has left us with complex and highly interdependent developmental pathways. Many of our anatomical peculiarities, such as the placement of our trachea in front of our esophagus, which leaves us vulnerable to choking, can be understood as the result of our evolutionary history—in this case, our history as aquatic organisms whose respiration depended on gills rather than lungs. The development of our respiratory and gastrointestinal systems is now so deeply embedded in the whole of our development that mutations which might have led to a safer anatomic design would almost certainly have been lethal (Held 2009). Moreover, because of our complex internal organization and our complex interactions with the external world, virtually every gene has multiple phenotypic consequences. Evolution frequently involves tradeoffs or compromises, such that natural selection leads to suites of traits that are not perfect or ideal, but work well enough for survival and reproduction, and are better than the available alternatives.

Finally, despite natural selection, survival and reproduction may be constrained by limitations of environmental resources, in the way originally envisioned by Malthus. Availability of nutritional resources is thought to have played a major role in evolution and nutritional deficiencies are still important causes of disease and death.

The remainder of this book will discuss these evolutionary determinants of disease in greater detail. Understanding the evolutionary reasons for our susceptibility to disease complements the traditional biomedical understanding of the etiology and pathogenesis of disease. Together, these two perspectives on health and disease, the ultimate and the proximate causes of disease, will help us understand why we get sick as well as how we get sick, and will provide insights into interventions that might reduce the burden of disease. First, however, we shall discuss human demography. As we shall see, demographic processes play a central role in evolution.

2

Human demography, history, and disease

2.1 Introduction

Evolutionary fitness entails both survival and fertility. Organisms must survive to the age of reproductive maturity and must then be fertile and produce offspring. Differences in the fitness of organisms of different genotypes are the result of differences in these demographic parameters. In other words, genes affect fitness by influencing survival or fertility rates—or more precisely, because genes may have different effects at different ages, by influencing age-specific survival or fertility rates. The age-specific survival and fertility rates of populations, together with migration rates, determine the growth rates of these populations.

Age-specific survival and fertility rates also determine the age structure of populations. Because people's age affects their risk of acquiring specific diseases, survival and fertility rates affect the patterns of disease in populations. Conversely, diseases have demographic and evolutionary consequences because they affect age-specific survival and fertility rates. Diseases that increase mortality or decrease fertility will decrease population growth rates and reduce the size and density of populations. And diseases cause evolutionary changes in populations because they differentially affect the survival and fertility of people with different genotypes. To understand the state of the human population and our burden of disease today, we need to understand the principles of demography and population growth, and appreciate the demographic and disease history of the human population.

2.2 Population growth: birth rates and death rates

In 2011, the human population was thought to have exceeded 7 billion people. According to World Health Organization (WHO) and official governmental estimates, there were roughly 133 million live births and 56 million deaths in 2011, resulting in an increase of about 77 million people in the world population that year (Central Intelligence Agency 2012). Birth and death rates are often expressed as events per 1000 people per year. In these terms, the birth rate was about 19/1000 (133 million/7 billion), the death rate was roughly 8/1000, and the annual growth rate was approximately 11/1000, or 0.011. One widely used measure of population growth rates is the doubling time, the time that would be required for the population to

Evolution and Medicine. First Edition. Robert L. Perlman © Robert L. Perlman 2013.
Published 2013 by Oxford University Press.

double in size *if* it maintained its current annual growth rate. When the growth rate is slow, population growth can be modeled as exponential growth. Under these conditions, the population doubling time equals ln2/annual growth rate, where ln2 is the natural logarithm of 2, or approximately 0.693 (Cohen 1995). If the human population continued to grow with an annual growth rate of 0.011, it would double in just over 60 years (0.693/0.011 = 63). Birth and death rates are constantly changing, however, and so the doubling time provides only a measure of population growth at the moment; it is not meant to be a prediction that the world population will actually double in 63 years.

These worldwide averages conceal large variations in birth and death rates among countries. Birth rates range from 7–8/1000 in Monaco, Hong Kong, and Singapore to 50/1000 in Niger, and death rates vary from less than 3/1000 in several countries in Western Asia (Jordan, Bahrain, and Kuwait, and some neighboring countries) to over 17/1000 in South Africa. These so-called crude death rates depend on the age structure of the population and so provide only limited information about the relative health of populations. Countries with relatively young populations, such as those in Western Asia, have lower crude death rates than do economically developed countries with older populations, even though these "young" countries have higher infant mortality rates and lower life expectancies.

The U.S. population in 2011 was approximately 315 million. Birth and death rates in the United States were about 14/1000 and 8/1000, respectively (roughly 4.3 million births and 2.5 million deaths), resulting in an endogenous or natural growth rate of about 6/1000, or 1.8 million people a year. In addition, the United States accepts a little over 1 million legal immigrants per year, so that the official population growth in the United States in 2011 was close to 3 million people, or 9/1000.

Let's go back to the roughly 56 million deaths in the world per year. To what are these deaths due? According to official estimates, several classes of diseases together account for more than half of these deaths. Cardiovascular diseases, primarily ischemic heart disease and cerebrovascular disease (heart attacks and strokes), are major causes of death, accounting for perhaps 17 million deaths per year. Infectious and parasitic diseases cause some 9–10 million deaths per year. Of these, the most important are respiratory and diarrheal diseases, which are caused by many different pathogens, HIV, tuberculosis, and malaria. Finally, cancers of all sorts are responsible for another 7–8 million deaths per year. WHO statistics underestimate the contribution of malnutrition to death rates because malnourished people typically die of infectious diseases before they actually starve to death. Maternal and child undernutrition is thought to be the underlying cause of death of some 3.5 million children under age five annually (Black et al. 2008). Most of these deaths would be reported by the WHO as due to infectious diseases. Diseases are not the only causes of death. Roughly 6 million people die each year of automobile and other accidents, violence, suicide, and war. This medical classification identifies proximate causes of death. Later, we shall discuss the ultimate or evolutionary causes of these diseases.

Deaths in infancy and childhood have greater evolutionary consequences than do deaths in post-reproductive years. In other words, differences in mortality rates of infants and children with different genotypes result in greater differences in fitness than do differences in mortality rates later in life. For the world as a whole, infant mortality, defined as deaths in the first year of life, is about 40/1000 live births. As with birth and death rates, there are large differences among countries. Infant mortality rates range from 2–6/1000 live births in most economically developed countries to over 100/1000 live births in some developing countries. Afghanistan, with an estimated rate of 122/1000 live births, currently has the highest infant mortality in the world. Infant mortality in the United States is the highest among developed countries at 6/1000, but this number too conceals huge disparities—infant mortality is about 5/1000 for whites and 12/1000 for blacks. In economically developed countries, relatively few children die between the first year of life and the time they reach sexual maturity. In contrast, in some developing countries, another 10–15% of children die before they reach the age of reproduction.

In developed countries, the major causes of infant mortality are congenital malformations and chromosomal abnormalities, complications arising from labor and delivery, the consequences of prematurity and low birth weight, sudden infant death, and accidents. Many of these deaths are thought to result from random errors in development that are not heritable. If this is true, natural selection may not be able to reduce infant mortality much below the low end of its current levels. Infant mortality due to developmental abnormalities represents the tail end of a large but largely unrecognized loss of fertilized zygotes and embryos in utero. The majority of fertilized zygotes do not implant in the uterus and are washed out with the mother's next menstrual period, and a large percentage of those that do implant fail to develop and are lost as spontaneous abortions (Macklon et al. 2002). Many of these embryos have developmental abnormalities and would probably not survive even if they were carried to term. We don't know if there are genes that underlie the detection and spontaneous abortion of embryos with chromosomal or developmental abnormalities. If there were, they would enhance fitness by reducing investment of maternal resources in embryos that are not likely to be viable (Stearns 2012).

In developing countries, most infant and childhood deaths are due to malnutrition and infectious diseases, especially respiratory and gastrointestinal infections. These deaths are not the inevitable result of developmental errors but are tragedies that can and should be reduced or eliminated by economic development and improvements in public health. The 20–25% of newborns in these countries who die before reaching reproductive maturity almost certainly do not represent a random genetic sample of the populations from which they come. Recall from Chapter 1 that natural selection is a process of nonrandom elimination. As long as these populations continue to suffer high infant and childhood mortality, they must be continuing to evolve. Genotypes associated with decreased infant and childhood mortality, most likely because of increased resistance to infectious diseases, must be increasing in frequency, while those associated with increased pre-reproductive deaths are decreasing.

2.3 Population growth in age-structured populations: fertility rates

Birth rates and death rates are population-wide demographic measures. We can also think about population growth in terms of the average fertility rates of individuals in the population.

Human populations are what demographers call age-structured. Populations are made up of individuals of different ages, whose survival (or mortality) and fertility rates depend on their current ages. Population pyramids provide a graphical illustration of the age structure of a population. These graphs show the number of males and females in different (typically five-year) age groups. Figure 2.1 shows population pyramids for Nigeria (2010) and the United States (2012). Nigeria has relatively high birth rates and death rates (40/1000 and 16/1000, respectively). Its population pyramid has a wide base (i.e. the country has many infants and young children) and actually resembles a pyramid; there are fewer and fewer people in each successive age group. In the United States, which has relatively low birth rates and death rates, the population pyramid is more columnar than pyramidal; there are roughly the same number of people in all age groups from 0–4 through 50–54.

Fertility rates are usually expressed in terms of female reproduction, if for no other reason than that mothers are present at the birth of their infants and information about mothers is likely to be more reliable than information about fathers. Even in the United States, which maintains relatively good records of vital statistics, information about the father's age is missing on some 15% of birth certificates and may be inaccurate in a significant percentage of the others. Indeed, we don't really know how often the person who is named as the father is in fact the genetic father of the newborn (Landry and Forrest 1995). Because fertility rates are not constant throughout adult life, they are often expressed as age-specific fertility rates, that is, as the fertility rates of women of a given (again, typically five-year) age group. Figure 2.2 presents age-specific fertility rates for the U.S. female population in 2008 (Hamilton et al. 2010). Fertility rates are very low in girls aged 14 and under, are maximal in 25–29-year-old women, and decline back to low levels in women over 45.

The average number of children that would be borne by a sexually mature woman throughout her reproductive lifetime if her fertility conformed to the current age-specific fertility rates in the population is known as the total fertility rate, or TFR. TFR is the sum of the age-specific fertility rates. The TFR of American women in 2008 was 2.09. TFRs in other countries range from about 1 in Hong Kong, Singapore, and Taiwan to over 7 in Niger; for the world as a whole, the TFR is about 2.5. TFR is an important determinant of population growth or decline. For a population to remain nongrowing or stationary, each adult, sexually mature woman must on average give birth to one daughter who herself survives to reproductive maturity (roughly, say, age 15). Equivalently, each newborn girl must on average produce one newborn daughter during her lifetime. (Demographic terminology may be confusing. A stationary population is a nongrowing population, while a stable population is a population with a stable age structure. Stable populations are not necessarily stationary.) The average number of adult daughters produced by a cohort of adult women depends on the number of children they give birth to during their reproductive lifetime—that is, their TFR—the proportion of these children who are daughters, and the probability that their newborn daughters will survive to the age of sexual maturity. In other words, the average number of adult daughters produced per woman equals:

(TFR) × (probability of daughter) × (probability of survival to adulthood)

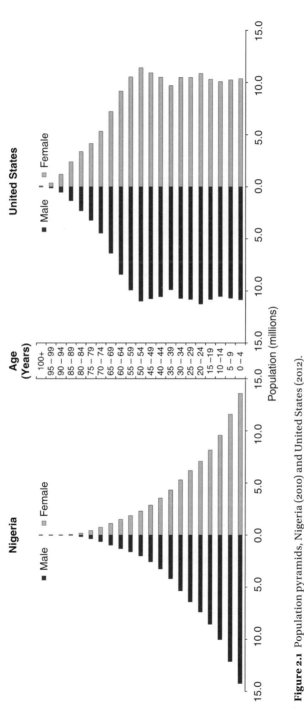

Figure 2.1 Population pyramids, Nigeria (2010) and United States (2012).

Source: U.S. Census Bureau, International Data Base <http://www.census.gov/population/international/data/idb/informationGateway.php.>

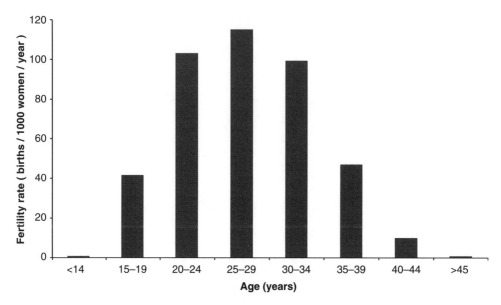

Figure 2.2 Age-specific fertility rates, United States, 2008.

Source: Hamilton et al. 2010.

Leaving aside migration, in a stationary (nongrowing) population, each woman must on average replace herself with one daughter who survives to sexual maturity. The replacement rate is the TFR required to maintain a stationary population. If the TFR is greater than the replacement rate, the population will grow; if it is maintained at less than the replacement rate, the population will decline.

The sex ratio at birth is slightly skewed towards boys (Arbuthnott 1710–1711). The ratio of newborn boys to girls differs in different human populations but is close to 1.05:1 (approximately 105 boys for every 100 girls) so that the proportion of newborns that are girls is roughly 0.49. Our male-biased sex ratio is presumably an evolutionary response to the increased mortality of adolescent boys and young adult men, which is due in large part to sexual selection for aggressive or risk-taking behavior (Arbuthnott 1710–1711; Kruger and Nesse 2004).

In the United States and most other economically developed countries with low infant mortality rates, the life expectancy of a newborn girl is greater than 75 years and the probability that a baby girl will survive to sexual maturity is about 0.97. In developed countries, then, the replacement rate—the TFR required to maintain a stationary population—is 1/(0.49)(0.97) or about 2.1. In developing countries, where infant mortality rates are greater and the probability of survival to adulthood is lower, replacement rates are larger. The replacement rate for the world population today is on the order of 2.2. The global TFR of 2.5 is considerably greater than the replacement rate, which explains why the world population is growing.

Because of the future reproductive potential of the young people in a population, a growing population has "momentum" and will continue to grow for some time even after the TFR declines to the replacement rate. Although the TFR in the United States is very close to the replacement rate, the U.S. population is still growing and will continue to grow even in the absence of immigration. The population will stop growing only after its TFR remains at the replacement rate for several generations and the age structure of the population becomes stable. Despite this complication, estimates of TFR and replacement rates are still useful demographic indices. Like population growth rates, they provide snapshots of population trends rather than predictions.

From a population perspective, then, growth rates are determined by the balance between birth and death rates. If birth rates are greater than death rates, the population will grow. In terms of women's reproductive histories, population growth rates are determined by the relationship between TFR and replacement rate. If TFR is greater than the replacement rate, the population will grow. In stationary populations, birth rates equal death rates, but stationary populations may have high birth and death rates (perhaps 40–50/1000) or low birth and death rates (on the order of 10/1000). Likewise, women in stationary populations may have high TFR (in the range of six to eight) if pre-reproductive mortality is high, or low TFR (near two) if pre-reproductive mortality is low.

2.4 Age-specific death rates

Advances in medicine and public health over the last several hundred years, especially improvements in the quality of food and water supplies and the development of vaccines, have greatly reduced infant and childhood mortality in the United States and other economically developed countries. As a result, we have become an aging society. A few statistics will illustrate the changes in the age structure of the U.S. population over the last century. In 1900, life expectancy at birth (in the handful of states that reported vital statistics) was 47, the median age of the population was 23, and 4% of the population was over 65. In 1950, life expectancy was 68, the median age was 30, and 8% of the population was over 65. By 2000, life expectancy at birth had increased to 77, the median age of the U.S. population was 35, and 12% of the population was over 65 (Hobbs and Stoops 2002).

The aging of individuals is accompanied by an increased probability of their death, and so populations exhibit age-dependent increases in death rates. Age-specific death rates are commonly expressed as deaths per 100 000 people of a given age or age range. Figure 2.3 shows current age-specific mortality rates for the U.S. population. Mortality rates are high in the newborn period (recall that the infant mortality rate in the United States is about 6/1000 or 600/100 000), decline in childhood, reach a nadir at about age 9 or 10, and then increase progressively with increasing age. At all ages, death rates for males are higher than those for females. The death rate for boys and men increases sharply between the ages of 15 and 29; death rates for young men are two to three times higher than those for young women. As we discussed earlier, more boys are born than girls. Because of the increased mortality rates of boys and men, the numbers of men and women of reproductive age are approximately equal; then, after about age 55, women outnumber men. Between the ages of about 30 and 80, age-specific death rates for both men and women double roughly every 10 years. Finally, at very old ages, death rates seem to plateau and may even decline (not shown in figure).

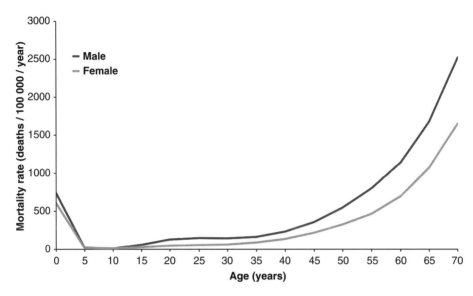

Figure 2.3 Age-specific mortality rates, male and female, United States, 2007.

Source: National Vital Statistics Reports, Vol. 59, No. 9, 28 September 2011.

The stabilization of mortality rates at old ages is not well understood but appears to be a real phenomenon in other species as well as in humans (Rose et al. 2006).

Age-specific mortality rates can be used to estimate life expectancy, the average number of years a population cohort would be expected to live if it experienced the current age-specific death rates in their populations. In 2011, life expectancy at birth in the United States was about 78 years (up from 77 in 2000), a little more for women, a little less for men (Central Intelligence Agency 2012). Life expectancy at birth ranged from less than 50 in Afghanistan and several African countries to almost 90 in Monaco. For the world population as a whole, it was about 68 years. Life expectancy is a good measure of the health of a population but, like population growth rates or TFR, it can't be used in a predictive sense because age-specific death rates are constantly changing. If mortality rates continue to decline, the average lifetime of babies born in the United States today will be greater than 78 years.

Age-specific mortality rates can also be used to construct survivorship curves, which show the percentage of a population cohort that is expected to survive to a given age, again assuming that this cohort experiences the current age-specific mortality rates in their populations. Figure 2.4 shows a survivorship curve for the U.S. population. Median survival or median life span is about 81 years. Life expectancy is an estimate of the average number of years a population cohort will live, not an estimate of the median age of death. Because life expectancy is heavily weighted by infant and childhood mortality rates, median survival is higher than life expectancy. Together with age-specific fertility rates, survivorship curves provide estimates of the expected future reproduction of an age cohort. As we shall see, the expected future reproduction of an age cohort plays an important role in the evolutionary theory of aging.

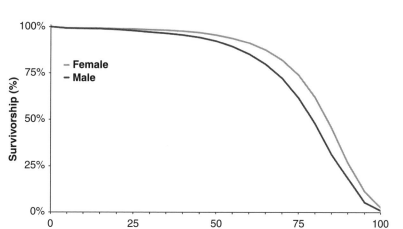

Figure 2.4 Survivorship curves, male and female, United States, 2007.

Source: National Vital Statistics Reports, Vol. 59, No. 9, 28 September 2011.

Maximum life span, the maximum age to which individuals in a species can live, is a less meaningful statistic than life expectancy or median survival, because it measures one extreme of the distribution of age at death and provides little information about the population as a whole; it is used primarily in comparative studies of aging. The human life span is conventionally taken as 120 years, probably because Moses was said to have lived that long. The oldest person whose birth and death are well documented was Jeanne Calment, a French woman who died at age 122. But maximum life span is the demographic equivalent of home run records in baseball. Just as baseball records are broken, in the future some people will undoubtedly live longer than 120 years.

2.5 History of human population growth

This, then, is a picture of the current state of the human population. What do we know about population growth and disease during human history? In *How Many People Can the Earth Support?*, the demographer Joel Cohen has integrated a wealth of data to estimate the history of the human population (Cohen 1995). Likewise, in *The Origins of Human Disease*, the medical historian Thomas McKeown has tried to identify the major causes of death throughout human history (McKeown 1988). Both of these works contain considerable conjecture, as we do not have good data about population growth or causes of death until the last several hundred years, and some of the authors' conclusions are controversial. Nonetheless, these summaries are based on the best available information and together they provide a plausible and coherent picture of the history of the human population.

Currently available evidence suggests that our species, *Homo sapiens*, arose in East Africa, in or near what is now Ethiopia, some 150–200 000 years ago, and began expanding out of Africa some time after about 100 000 years ago (Tattersall 2009). Throughout most of human history, the human population was essentially stationary. Until the agricultural revolution, which occurred independently in several parts of the world beginning roughly 12 000 years ago, population growth rates were close to zero, with doubling times greater than 10 000 years (Cohen 1995). Population growth was probably limited by the availability of food, by proximate mechanisms that included anovulation and infanticide or infant abandonment, as well as frank starvation. Changes in the abundance of food must have led to short term, local changes in population size. Over the long term, however, the population density in foraging societies was likely to have been more or less constant; population growth was due primarily to the geographic spread of the human population. Food shortages were probably major stimuli for population movements throughout human history, as they are even today. People in foraging societies also suffered (and continue to suffer) from infectious or parasitic diseases. The pathogens that caused these diseases had particular characteristics, such as the ability to cause long-lasting infections in humans, to infect other species, or survive for long periods in soil or water, that enabled them to be maintained in sparse and mobile populations (Diamond 1997). Parasitic worms, or helminths, were especially important pathogens throughout our evolutionary history. Birth rates and death rates during most of human history were roughly equal, probably on the order of 40–50/1000. TFR was probably between four and eight—each adult woman had on average four to eight children, or two to four daughters—but only a quarter to half of newborns survived to adulthood; in other words, TFR was close to the replacement rate. Again, these are long term averages that must conceal large local and short term fluctuations. Although an infant and childhood mortality rate of 50–75% strikes us as appalling, it is low in comparison to the pre-reproductive mortality of many other mammalian species. And recall that pre-reproductive mortality rates in many countries today are still on the order of 20–25%.

The transition from foraging to agriculture permitted an increase in population growth rates. Agriculture spread primarily because agriculturalists out-reproduced foragers, not because foragers adopted an agricultural lifestyle. Even after the agricultural revolution, population growth remained slow, with doubling times measured in thousands of years. Paradoxically, although the growth rates of agricultural populations were greater than those of foragers, agricultural people were probably less well nourished and less healthy than were people in foraging societies. Agriculturalists continued to suffer from undernutrition. Their fertility may have increased as a consequence of their decreased mobility or decreased workload. Just as natural selection optimizes fitness rather than health, cultural practices such as farming may spread because they increase fitness, even at the expense of health (Lambert 2009).

Beginning with the agricultural revolution and accelerating with the development of urban centers some 6–7000 years ago, infectious or parasitic diseases became more important causes of death, and especially of infant and childhood deaths. Several factors led to the rise of infectious diseases. First, as people domesticated animals, they began to live in close proximity with these animals. Many infectious diseases, especially epidemic diseases such as measles and smallpox, are zoonoses—that is, they are caused by organisms that were originally transmitted to humans from our domesticated animals (Diamond 1997). In addition, dense urban population centers supported pathogens that could not be maintained in smaller populations, and fixed settlements were conducive to fecal-oral and water-borne transmission of these pathogens. Infectious diseases restrained population growth even as food became more plentiful.

Only in the last few centuries have population growth rates begun to increase rapidly. The industrial revolution, which was accompanied by European exploration and colonization of much of the rest of the world, led to what has been called the global agricultural revolution, the worldwide exchange of domesticated plants and animals (Crosby 1971). This, together with more efficient use of farmland, increased urbanization, and better transportation, led to great increases in food production and distribution. By the end of the eighteenth century, the world population was growing with a doubling time on the order of 100 years. The increases in population growth rates since the industrial revolution were probably due both to a decrease in infant and childhood mortality and to an increase in fertility resulting from better nutrition (Frisch 2002). Fueled by the increased population density that accompanied urbanization, however, epidemic infectious diseases continued to be major causes of death.

Later, advances in public health, including preventive measures such as vaccination and improvements in water supplies, sanitation, and food storage reduced the toll of infectious diseases and further reduced mortality rates, which led to even greater rates of population growth. By the middle of the twentieth century, annual population growth rates reached a maximum of about 2% or 0.02, corresponding to a global population doubling time of about 35 years (Cohen 1995). It is ironic that while Malthus (1798) probably provided an accurate description of the human population until about the time he wrote, he did not anticipate the ways in which the industrial revolution and its aftermath would increase food production and distribution, and would support previously unimagined increases in population growth.

What Cohen calls the fertility revolution began in Western Europe in the late eighteenth century and has since spread to the United States and other economically developed countries (Cohen 1995). The fertility revolution is part of a broader demographic change known as the demographic transition. The demographic transition is a model that describes the way societies have passed, or are passing, from a regime characterized by high birth and death rates and low population growth, through a transitional period in which death rates decline but birth rates remain high and so population growth increases, to a state in which birth rates decline to approximate death rates and population growth rates decrease towards zero. We have seen how improvements in food production, distribution, and storage, together with advances in public health, decreased death rates and increased population growth. The fertility revolution is characterized by decreases in fertility rates and in population birth rates, as women have consciously or unconsciously chosen to postpone and limit their reproduction. In most countries, birth rates did not decline until several generations after the fall in death rates, and women in higher socioeconomic classes decreased their fertility before poorer and less well educated women did. As birth rates have fallen, population growth rates have decreased. As we noted earlier, the world population today has an annual growth rate of about 0.11, which corresponds to a doubling time of a little over 60 years. Figure 2.5 shows an idealized demographic transition, as a stationary or slow-growing population with high birth and death rates evolves into a stationary or slow-growing population with low birth and death rates. In many countries, TFR is now lower than the replacement rate; if TFRs remain low (and in the absence of immigration), the populations in these countries will eventually begin to decline.

The demographic transition has puzzled evolutionary biologists because it seems to contradict a principal tenet of the theory of evolution: that biological species, including humans, have evolved in ways that maximize their reproductive fitness. Biologists and social scientists have advanced a number of hypotheses to account for this apparent contradiction (Borgerhoff Mulder 1998). Some researchers have suggested that reduced fertility may actually be a reproductive strategy that enhances long term reproductive success. The best current understanding of the demographic transition, however, is that the decision to limit fertility does not increase fitness and from a purely evolutionary perspective it is indeed maladaptive.

Recall that fitness entails more than reproducing. It involves helping children grow and develop to the point that they can themselves reproduce. We must balance the number of children we have with the "quality" of our children, the likelihood that they will survive to sexual maturity and will then be fertile and reproduce. We can increase the chances of our children's survival by committing resources to them, a process referred to as parental investment. Because of resource limitations, there is often a tradeoff between the quantity and quality of children. Breastfeeding is a good example of parental investment and of the quantity/quality tradeoff. Breastfeeding provides resources to our children that promote their growth, development, and survival. But breastfeeding is accompanied by temporary anovu-

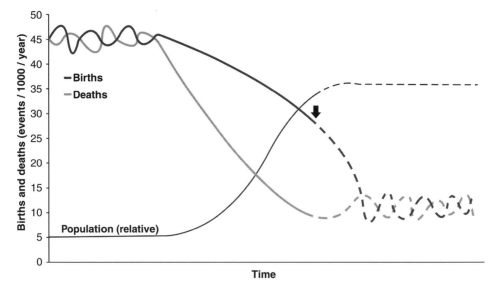

Figure 2.5 The demographic transition. The arrow indicates global birth and death rates today. The dashed lines are future projections.

lation and infertility; it reduces reproduction but enhances fitness. Our ancestors presumably evolved psychological as well as physiological mechanisms that enabled them to adjust parental investment and reproduction in ways that optimized their fitness.

For most of human evolution, at least until the onset of the agricultural revolution, wealth was primarily embodied and social wealth, in the form of skills and social relationships that helped our ancestors acquire food and find mating partners (Kaplan et al. 2002). In this environment, wealth was closely correlated with fitness. Skill at acquiring food led to better nutrition, which enhanced fertility and, perhaps more importantly, reduced the risk of infant and childhood mortality. Moreover, because we evolved a lifestyle that was dependent on hard-to-acquire foods such as meat, roots, nuts, and seeds, parental investment in children, in helping them acquire the skills and social relationships they needed to become successful adults, enhanced their fitness (Kaplan et al. 2003). Because of our evolutionary history, we may well have evolved psychological predispositions to pursue socioeconomic success and to invest heavily in our children, even if that means limiting reproduction.

We still seek socioeconomic success and, because we recognize that raising children is expensive, we limit our fertility and invest considerable resources into our children. Today, however, wealth is primarily extra-somatic, in the form of money, land, and other material possessions. Limited reproduction together with high parental investment helps children become economically successful and it preserves or accentuates socioeconomic disparities. In societies where infant and childhood mortality is already low, however, wealth no longer translates into reproductive success. In these societies, limitation of fertility decreases reproductive fitness; in evolutionary terms, it is maladaptive. This behavior apparently results from a mismatch between our evolved psychological biases and our contemporary environment (Goodman et al. 2012). As we shall discuss in Chapter 11, traits that were preserved by natural selection because they were beneficial in ancestral environments may become maladaptive (again, from an evolutionary perspective) as a result of environmental changes.

Many factors may contribute to women's or couples' decisions to limit their fertility. The decline in infant and childhood mortality has reduced the desire for "replacement children" who might previously have been needed to contribute to the family economy or take care of their parents in their old age. Urbanization has decreased the economic benefits of children to their families and has increased the cost of child rearing. Cultural norms in many social groups encourage small families. Decreasing fertility appears to be closely correlated with the availability of educational and career opportunities for women. Many women who have these opportunities choose to pursue them, because they can achieve greater economic success and may also achieve greater personal rewards by working outside the home. In general, educational and career opportunities for women have increased as societies have become more economically developed. As economic development and opportunities for women spread in countries that are currently underdeveloped, worldwide fertility rates may continue to fall. But the model of the demographic transition was developed to describe demographic changes in Western populations; how well this model will fit developing countries remains to be seen. Cultural traditions and values in some societies may hinder or prevent the spread of educational and career opportunities for women, and may encourage women to forgo these new lifestyles and maintain high fertility rates.

The declines in birth and death rates that characterize the demographic transition have been accompanied by an epidemiologic transition (Omran 1971). As we have seen, the major causes of death are changing from infectious diseases and famine to cardiovascular diseases and cancer. McKeown called these new diseases "diseases of affluence" (McKeown 1988). This term is a misnomer, however, because these diseases are not restricted to the affluent; they afflict poor people and are becoming increasingly prevalent in developing countries. "Western diseases" is problematic for the same reason. A more descriptive term is "degenerative and man-made diseases," which conveys the understanding that the prevalence of these diseases is due in part to aging of the human population and in part to man-made changes in the human environment (Omran 1971).

2.6 The future of the human population

Birth rates are decreasing and population growth is slowing. But we can't foresee how rapidly the demographic transition will spread to developing countries that still have high birth rates and so we can't predict how the human population will grow (or decrease) in the future. If present trends continue, the global human population may return to a state in which birth rates will be very close to death rates and in which population growth rates will again be negligible. The near-stationary populations of the future will differ from those that existed during most of human evolution because instead of a TFR of four to eight and a pre-reproductive mortality rate of 50–75%, we will have a global TFR a little above two and an infant and childhood mortality rate of 1% or less. In stationary populations, the birth and death rates equal the reciprocal of the life expectancy (Coale 1974). In a stationary population with a life expectancy of 20 years, for example, 1/20 of the population, or 50 people per thousand, die each year and an equal number are born. In the near-stationary populations of the past, life expectancies were probably on the order of 20–25 years, which would correspond to birth and death rates of 40–50/1000. Future populations might have life expectancies on the order of 80 to 90, with birth and death rates in the range of 11–12/1000. It is difficult to predict how long it will take the human population to reach this new stationary state (if we ever reach it) or how large the population will be when we do.

As the demographic transition proceeds, the changes in the burden of disease that characterize the epidemiologic transition will become more pronounced. Again assuming that present trends continue, infectious diseases and famine will become less important, while degenerative and man-made diseases will become more prominent. But the age structure of a population is an important determinant of economic growth as well as of the burden of disease. As the population ages, the proportion of people who are economically productive may decline. The resulting decrease in economic growth rates may also have adverse health consequences.

As we discussed above, even in the absence of any interventions by parents, birth rates for boys are slightly higher than those for girls. Some societies have had and continue to have a cultural preference for producing and raising sons rather than daughters, and the sex ratio in these societies has been further skewed by the selective abandonment or infanticide of new-born girls. A decrease in fertility rates, together with the availability of prenatal sex determination and sex-selective abortion, is leading to markedly unbalanced sex ratios in some countries. For example, some provinces in China have sex ratios at birth of over 130 boys per 100 girls. In the country as a whole, more than a million more boys than girls reportedly are born per year. China is now estimated to have an excess of 32 million males under the age of 20 (Zhu et al. 2009). The consequences of this unprecedented male excess are hard to predict but are likely to be significant.

3
Evolutionary genetics

3.1 Introduction

When Gregor Mendel (1865) carried out his research on the mechanisms of heredity, he chose
to study characters that existed in two clearly distinct forms. For example, he studied plants
that had either tall (6 to 7 feet) or short (¾ to 1½ feet) stems. He considered the variations
within the two groups to be unimportant (Mendel 1865). After Mendel's experiments became
widely known and genetics had become an active field of research, the large phenotypic
effects of the mutations studied by geneticists seemed incompatible with the graded, continuous variation that was at the heart of Darwinian evolution. This led to a prolonged and
heated controversy between "Mendelians" and "Darwinians" about the mechanisms of evolutionary change. It was not until the 1930s and 1940s, in what has been called the modern
synthesis, that evolutionary biology was integrated with genetics (Huxley 1942). Since that
time, population genetics has played a central role in our understanding of evolution (Hartl
and Clark 2006). And as we are learning more and more about the human genome, population genetics is increasingly informing our understanding of disease and the practice of
medicine.

The haploid (single copy, as found in sperm or oocytes) human genome comprises
approximately 3 billion nucleotides and contains roughly 21 000 protein-coding genes,
located at specific places, or loci, on one of our 23 chromosomes. Only a small fraction, probably less than 2%, of the nucleotides in our genomes code for protein sequences; the overwhelming majority of our genomic DNA has other functions. The ENCODE (Encyclopedia of
DNA Elements) project, which is generating an almost overwhelming amount of data, will
undoubtedly revolutionize our understanding of the organization and function of the human
genome (Stamatoyannopoulos 2012). The ENCODE project has identified thousands of DNA
sequences, accounting for over one-half of our genomic DNA, that are transcribed into RNA
molecules that do not code for proteins but have other biological functions (Esteller 2011).
There are probably as many of these noncoding RNA genes as there are protein-coding
genes. Noncoding RNAs include ribosomal RNA and transfer RNAs, which are essential for
protein synthesis, and a large number of RNA species that bind specifically to complementary nucleotide sequences in DNA or messenger RNA and participate in the regulation of
transcription, RNA processing and turnover, or translation. Other DNA sequences are regulatory sequences that control the expression of the protein-coding structural genes,
sequences that help maintain chromosome structure, repeating sequences that have

Evolution and Medicine. First Edition. Robert L. Perlman © Robert L. Perlman 2013.
Published 2013 by Oxford University Press.

replicated and spread in our genomes, or pseudogenes, historical relics of once-active genes that have been rendered nonfunctional by mutation but have not been eliminated from our genomes.

Most genes are polymorphic; that is, they exist in multiple, alternate forms, or alleles, within the gene pool of a population. Alleles differ in DNA sequence and so may encode proteins or noncoding RNAs with different sequences and different biological activities. As a result, organisms that have these different alleles may develop different phenotypes. Geneticists often think of evolution as a change in the frequencies of alleles or genotypes, and their associated phenotypes, in a population over time.

An early criticism of Darwin's theory of evolution was the concern that, because natural selection continually removes the less-fit variants from a population, it would decrease or eliminate the heritable variation that is required for evolution. In today's language, wouldn't natural selection lead to the spread and ultimately the fixation of the single genotype that had the highest fitness? Natural selection does act to decrease genetic diversity within populations. As we shall see, however, a number of other processes maintain the genetic variation on which evolution depends. Some of the early publicity surrounding the Human Genome Project may have revived the idea, reminiscent of the essentialist view of biology we discussed in Chapter 1, that there is a single normal or ideal human genome. But our species, like other species, is characterized by abundant genetic and phenotypic diversity—indeed, this diversity is one of the properties that distinguishes species from clones. Each of us has a unique genome, with a unique combination of protein-coding and noncoding DNA sequences. Many alleles are maintained in our genomes even though they increase the risk of disease and so may decrease reproductive fitness. The study of evolutionary medicine involves understanding the reasons for the persistence of these disease-associated alleles and their significance for the people who carry them.

3.2 Other evolutionary processes: mutation, genetic drift, and migration

Although biologists often talk about Darwinian theory as the theory of evolution by natural selection, natural selection is not the only evolutionary process. Mutation, genetic drift, and migration (gene flow) can also lead to changes in allele or genotype frequencies in populations over time. New alleles arise by mutation. Mutations were initially recognized as heritable factors that produced altered phenotypes. After genes were defined as DNA sequences that specified the amino acid sequences in proteins or the nucleotide sequences in noncoding RNAs, mutations were understood as changes in these DNA sequences, due to errors in DNA replication, which are transmitted from parents to offspring. Most mutations are single nucleotide changes, the replacement of one nucleotide by another at a specific site in the genome, leading to single nucleotide polymorphisms. But mutations may also involve larger genetic changes, including insertions and deletions of one or more nucleotides, duplications of DNA sequences (resulting in copy number variations), rearrangements and translocations of large portions of chromosomes, and even duplication or loss of entire chromosomes.

Although mutations are random changes in DNA sequences, mutation rates are themselves subject to natural selection and to physiological regulation. Mutation rates are the net result of the chemical stability of DNA, the fidelity of DNA synthesis, and the activities of DNA repair processes.

The best current estimates of the mutation rate in humans are on the order of 12–15×10^{-9} per nucleotide site per generation (Lynch 2010; Scally and Durbin 2012). By far the most common mutation in humans is the replacement of cytosine by thymine, abbreviated as a C→T transition, occurring especially at CpG sequences, dinucleotides where cytosine is followed by guanine. The C→T transition can serve as a model for understanding mutations in general. Cytosine residues in DNA are subject to spontaneous oxidative deamination, which results in the formation of uracil. We have enzymes that recognize uracil as an abnormal base and excise it from DNA. Other enzymes repair DNA sequences that have missing bases. These enzymes use the complementary DNA strand as a template. Since the complementary strand contains guanine at this position, the repair enzymes regenerate the original cytosine-containing sequence. Cytosine deamination results in a mutation only if DNA replication occurs before the original sequence is restored; in that case, the presence of uracil will lead to the insertion of adenine instead of guanine on the newly synthesized strand. If cytosine residues in DNA are methylated, however, their deamination produces thymine, which is not an abnormal base and so is not removed by base excision repair enzymes. As we shall discuss later, DNA methylation (and especially methylation of cytosine residues in CpG dinucleotides) is an important mechanism for the regulation of gene expression. Mutations caused by C→T transitions are evidently one price we pay for the benefits of this regulatory mechanism.

Given our haploid genome of roughly 3×10^9 nucleotides, a mutation rate of 12–15×10^{-9} per nucleotide site will result in something like 35–50 new mutations in each germ cell or 70–100 in each new zygote. The fate of new mutations depends on their phenotypic effects and also on chance. Mutations that enhance fitness will tend to spread in a population, while those that decrease fitness will tend to decrease in frequency or be eliminated. Again, this is natural selection. Even though only a small fraction of new mutations are deleterious, new mutations are a major reason for the loss of zygotes prior to implantation and for spontaneous abortions (Macklon et al. 2002), and are also an important cause of disease (Crow 1997).

Genetic drift plays a major role in determining the fate of new mutations. Recall that genetic drift refers to changes in allele frequencies due to random sampling in the transmission of alleles from one generation to the next. Most species have high juvenile mortality rates. The majority of new mutations are lost simply because the organisms that carry them die before reaching reproductive maturity. If these individuals do survive and reproduce, the random transmission of alleles from parents to offspring during meiosis and sexual reproduction can lead to the loss of new mutations, even if they are beneficial, and to the spread of deleterious ones. Most single nucleotide changes, including most C→T transitions, are neutral—that is, they have no significant effect on evolutionary fitness. Most mutations occur in DNA sequences that do not code for proteins. Neutral mutations in protein-coding sequences may be synonymous mutations (changes in the nucleotide sequence that do not change the amino acid sequence of the protein for which the gene codes) or nonsynony-mous mutations in which the altered amino acid sequence does not significantly affect the biological activity of the protein. The fate of these neutral mutations is completely depend-ent on genetic drift.

Genetic drift is an especially important evolutionary process in small populations. During the geographic expansion of the human species, many populations were begun by small groups of colonizers or founders. The founders of these new populations possessed only a sample of the alleles in the populations from which they came. As a result, these new populations have less genetic diversity than did their parental populations. But while some alleles may be lost in the process of founding new colonies, alleles that by chance were present in the founders may remain at high frequencies in the new populations simply as a result of this so-called "founder effect" rather than because of selection. The pattern of genetic diseases among French Canadians in Quebec illustrates the medical importance of founder effects (Laberge et al. 2005). The present Quebecois population of about 6 million people is descended from some 8500 French settlers who came to Canada in the seventeenth and eighteenth centuries. Several dozen genetic diseases are present in unusually high frequency or have distinctive features in the Quebecois population. The incidence of these diseases reflects the frequency of disease-associated alleles that happened by chance to be present in the founder population.

Ancestral human populations are thought to have experienced bottlenecks, times when their population size was greatly reduced by disease or lack of food. Because survivors of bot-tlenecks, like the founders of new populations, contain only a sample of the alleles that were present in the parent population, bottlenecks also decrease genetic diversity and produce changes in allele frequencies in the resulting populations.

Recall that biologists commonly think of species as comprising populations of organisms that can interbreed in nature. Often, these populations are geographically isolated from one another. The organisms in geographically isolated populations may come to differ pheno-typically from those in other populations because they are developing in different environ-ments. In addition, geographic isolation or environmental separation leads to reproductive isolation, because organisms are more likely to mate with others in their vicinity than with those that are farther away. New mutations, natural selection acting on organisms in different environments, and genetic drift may then lead to genetic differences among populations. For the most part, these genetic differences are differences in allele frequencies. Reproductive isolation is rarely complete. Migration, or gene flow, refers to the movement of organisms and their genes between populations. Gene flow, which brings new alleles into populations and so reduces the genetic differences among them, has been an important evolutionary process in human history. Mutation and gene flow are the most important sources of genetic varia-tion in human populations. Occasionally, new alleles arise from the genetic recombination that occurs during sexual reproduction. And new genes may be introduced by viruses that become incorporated into the DNA of germ cells and are then transmitted to offspring. This process of "horizontal" gene transfer by viruses or other DNA elements is especially impor-tant in bacteria, which do not reproduce sexually. However genetic variations arise, their fate is determined by selection and genetic drift.

3.3 Genetic dominance

Given how well integrated our physiology and metabolism are, and how well adapted our evolutionary ancestors became to the habitats in which they lived (they were, after all, at least well enough adapted to have become our ancestors), it should not be surprising that new mutations which do have phenotypic effects are more likely to be harmful than beneficial. But deleterious mutations are not automatically and completely eliminated by natural selection. Many deleterious or lethal alleles are recessive—they produce their harmful phenotypic effects only in homozygous organisms, organisms that have two copies of the allele. These deleterious recessive alleles often encode a nonfunctional protein product. For many genes, the protein produced by the expression of one functional allele is enough to maintain a normal phenotype. In other words, alleles that encode functional protein products are often dominant. The reasons for the evolution of genetic dominance remain controversial. Dominance may have evolved as a result of selection, because beneficial dominant alleles will spread more quickly than will recessive alleles and because dominance may enhance fitness by providing a buffer that protects organisms from the deleterious effects of mutations (Fisher 1930). It seems more likely, however, that dominance is in general a nonselected outcome of the regulation of metabolic pathways or the structure of gene networks.

Many genes encode enzymes that are components of metabolic pathways. Changes in the activity of these enzymes typically have overt phenotypic effects only to the extent that they affect flux through these pathways. If a pathway comprises many individual enzymatic reactions, most of the enzymes will have only a small effect on flux through the pathway. Even loss of 50% of enzyme activity may have a negligible effect on the pathway (Kacser and Burns 1981). The genes that encode these enzymes will be dominant. Their dominance is a consequence of the properties and regulation of the metabolic pathways in which they participate (Fell 1997, pp. 106–7).

However it evolved, genetic dominance has important evolutionary consequences. When deleterious or lethal recessive alleles are rare, most copies of these alleles will be in heterozygous individuals and will be shielded from natural selection because they don't decrease the survival or fertility of the individuals who carry them. Only those relatively few alleles in homozygous individuals will not be transmitted to offspring and so will be removed from the gene pool of the population. These deleterious recessive alleles will increase in frequency in the population until the rate at which they are lost by natural selection equals the rate at which they are formed by new mutations. In technical terms, their steady state frequency will be determined by the balance between mutation and selection.

The steady state frequency of deleterious or lethal recessive alleles in randomly mating populations is roughly equal to the square root of the mutation rate. Before the advent of molecular genetics, when genes were thought of as indivisible units that existed in two allelic forms, mutation rates were estimated to be on the order of 10^{-5}–10^{-6} per genetic locus per generation, in which case the steady state frequencies of deleterious recessive alleles would be in the range of 10^{-3}, or 0.001. Genetic polymorphisms were conventionally defined as the presence of two alleles with frequencies greater than 0.01 because these frequencies were higher than could be accounted for by mutation-selection balance. Now we can distinguish alleles on the basis of their DNA sequences. Given a mutation rate of $12–15 \times 10^{-9}$ per nucleotide site per generation, mutation-selection balance should maintain deleterious recessive alleles at frequencies on the order of 10^{-4}. Today, a frequency of 0.01 or 0.001 is often used to distinguish "common" from "rare" allelic variants. The calculations are slightly different for dominant alleles and for sex-linked genes but the principle is the same. Natural selection may keep deleterious or disease-associated alleles at low frequencies but it cannot eliminate them completely (although, when they are rare, alleles may be lost by genetic drift).

Although most deleterious mutations are recessive, a few are dominant. These dominant negative mutations may be mutations in genes that encode subunits of larger proteins or protein complexes. Altered subunits may interfere with the assembly and function of these larger proteins. Dominant negative mutations exert their harmful effects in all the organisms that carry them. They will be more strongly selected against and so will be maintained at lower frequencies than deleterious recessive mutations.

3.4 Heterozygote advantage

Many mutations do not cause a loss of function. A mutant allele may encode a protein that has reduced biological activity. In other instances, a mutant allele encodes a protein that has a new or different activity. If the protein products of both alleles are biologically active in heterozygous individuals, the alleles are said to be codominant. The ABO blood group gene has three families of alleles, A, B, and O. All of the A alleles produce the A phenotype, while the B alleles produce the B phenotype; the alleles within each family can be distinguished only by their DNA sequences. Both A and B are dominant over O but the A and B families are codominant. Heterozygous people who have both an A allele and a B allele express both blood group antigens and have the AB blood group phenotype. Occasionally, fitness of the heterozygous genotype is greater than that of either homozygous genotype, resulting in what is known as heterozygote advantage. Because heterozygous individuals can pass on both alleles to their offspring, heterozygote advantage is one explanation for the maintenance of genetic polymorphisms.

Although heterozygote advantage can maintain genetic polymorphisms, it is often an unstable phenomenon, as there will always be selection for compensatory mutations that reduce the fitness disadvantage of the homozygous genotypes. Gene duplication is one mechanism that may overcome heterozygote advantage, because after a gene is duplicated every member of the population may have copies of both alleles. Color vision in humans and other Old World primates depends on at least two X chromosome-linked opsin genes, which encode proteins with sensitivities to long (red) and medium (green) wavelengths, and an autosomal gene that encodes a protein with sensitivity to short (blue) wavelengths. The multiple X chromosomal opsin genes apparently evolved from a single ancestral gene by gene duplication (Hunt et al. 2009). One plausible hypothesis is that a mutation in an ancestral red- or green-sensitive opsin gene gave heterozygous females a fitness advantage. Heterozygous females expressed both the red- and green-sensitive alleles and so had better red/green color discrimination and were better able to find food than were males or homozygous females, who had only one X-linked opsin allele. Gene duplication made it possible for the two alleles to spread and become fixed at different genetic loci, such that most members of the population then expressed both red- and green-sensitive opsin genes and had the fitness benefits of enhanced color discrimination. Additional rounds of gene duplication and gene loss have led to variations in the copy number of green-sensitive opsin genes in human populations. These additional opsin genes do not appear to be associated with changes in fitness. Their number is presumably fluctuating as a result of genetic drift.

3.5 Pleiotropy and epistasis

The traits that Mendel studied not only had two distinct and easily recognized forms, like long and short stems, but other characteristics that made his research so productive. In modern terms, Mendel examined traits in which 1) a single genetic locus makes a major contribution to the phenotype, 2) there appear to be only two alleles at each locus, 3) there is a close connection between genotype and phenotype, and 4) the genes have no obvious phenotypic effects other than the trait of interest (Mendel 1865). These are what we now think of as "simple" Mendelian traits. Whether it was good luck or good intuition that led Mendel to study these traits, his choice was fortunate, because it enabled him to interpret and make sense of his research, which ultimately led to the development and flourishing of the field of genetics. In the short run, however, Mendel's experiments may have led people to form misleading conceptions of the actions of genes and of the relationship between genotype and phenotype. We now know that many genes exist in far more than two allelic forms. Only rarely do genes have just a single effect on the phenotype of an organism. Most genes have multiple phenotypic effects; in technical terms, they are pleiotropic.

Conversely, most traits of biological or medical interest result from the interactions of several or many genes or gene products. The interaction of multiple genes in producing a trait is known as epistasis. Both pleiotropy and epistasis complicate the relationships between genotype and phenotype, or between genotype and the risk of disease. Alleles of pleiotropic genes may have both beneficial and deleterious effects. These alleles will increase in frequency until their deleterious effects on fitness (increasing the risk of disease, for example) balance their beneficial effects in the population as a whole. Moreover, because of epistatic interactions among genes, the fitness consequences of an allele at one locus may depend on alleles at other loci. Individual alleles may be beneficial in people with some genotypes and harmful in people with others. Again, these alleles will spread in a population until their beneficial or fitness-enhancing effects equal their harmful or fitness-reducing consequences. Pleiotropy and epistasis exemplify the tradeoffs that are at the heart of evolution.

Alleles of the α- and β-globin genes, which encode the α and β chains of hemoglobin, exhibit both pleiotropy and epistasis. Some mutations in these genes result in decreased synthesis of one or the other of the hemoglobin chains, causing a hemolytic anemia known as α- or β-thalassemia. But the thalassemia alleles are pleiotropic. As we discussed earlier, they increase resistance to malaria. These alleles have spread in regions where malaria was prevalent, particularly in the Mediterranean, because of heterozygote advantage. The fitness gain due to malaria resistance in these environments balanced the fitness loss due to thalassemia. The anemia seen in patients with thalassemia results in part from the unbalanced synthesis of the α- and β-hemoglobin chains. As a result, α- and β-thalassemia alleles are epistatic to one another. α-thalassemia alleles are harmful in people who have the normal β-globin allele because they cause thalassemia but they are beneficial in people who have β-thalassemia alleles because they reduce the imbalance between the two hemoglobin chains and decrease the severity of the disease. Because of these epistatic interactions, α- and β-thalassemia alleles are maintained in the same populations (Penman et al. 2009).

3.6 Linkage and hitchhiking

Genes are located at specific sites or loci on chromosomes and are physically as well as biologically linked to nearby genes; they are parts of the same long DNA molecule. Crossing-over or genetic recombination, the physical exchange of regions of DNA between homologous chromosomes during meiosis, is a relatively rare event; it occurs once or only a few times in each pair of chromosomes. Because of the low rate of genetic recombination, genes that are located close together on a chromosome may be transmitted from parents to children for many generations as a unit rather than separately. Until they are separated by recombination, these tightly linked genes form what is known as an extended haplotype block, which behaves as a single pleiotropic gene. Genetic linkage may cause deleterious alleles to increase in frequency if their harmful effects are outweighed by the benefits of a neighboring favorable allele. The major histocompatibility complex (MHC) is a set of genes that are important for the function of the immune system. The MHC locus is tightly linked to a gene involved in iron metabolism, some alleles of which can cause hemochromatosis, a disease caused by the accumulation of iron in the liver and other organs. One hypothesis for the high frequency of hemochromatosis in European populations is that the spread of a beneficial MHC allele resulted in the spread of a deleterious hemochromatosis allele that happened to be genetically linked to it (Distante et al. 2004).

3.7 Frequency dependent selection

In the case of heterozygote advantage, the fitness effect of an allele depends on the other allele at that genetic locus. When there are epistatic interactions among genes, the fitness effect of an allele depends on alleles at other genetic loci. The fitness effects of alleles may also depend on their frequency in the population. Some alleles increase fitness when they are rare but lose this fitness benefit as they become more common. This phenomenon, known as frequency dependent selection, is another mechanism that can maintain genetic polymorphisms in populations.

The British biologist R. A. Fisher suggested that frequency dependent selection could account for the near 1:1 sex ratio in humans and many other sexually reproducing species (Fisher 1930). Given the process of sexual reproduction, each baby has a mother and a father.

If there is an imbalance in the numbers of the two sexes, members of the less common sex will on average have more offspring than will members of the more abundant sex. Therefore, alleles that bias the sex ratio of offspring toward the less common sex will increase fitness. Parents who produce children of the less common sex will, on average, have more grandchildren. As these alleles spread in the population, the sex ratio will become closer to 1:1, and their fitness benefits will decrease. In this way, alleles that bias the sex ratio of offspring in both directions will maintain the ratio at close to 1:1. The sex ratio may also be maintained or manipulated by direct selection by parents. In some species, females are able to adjust the sex ratio of their offspring and so can preferentially produce offspring of the less abundant sex. Nonetheless, Fisher's proposal illustrates how frequency dependent selection might work. In this example, frequency dependent selection depends on interactions among members of a population, specifically, on sexual reproduction. In other instances, frequency dependent selection results from interactions among species. As we shall see, pathogens are an important cause of frequency dependent selection and so play a major role in the maintenance of polymorphisms in their host populations.

3.8 Epigenetic regulation of gene expression

The regulation of gene expression is a critical step in the relationship between genotype and phenotype. The development of phenotypes depends not merely on the presence of specific genes but on the expression of these genes and on the biological activities of the RNA or protein products for which they code. Regulation of gene expression is breathtakingly complex and can't be neatly categorized. Gene expression may be regulated in response to the physiological needs of the organism. These responses may be initiated by hormones or other regulatory molecules and are typically mediated by changes in the activity of transcription factors, proteins that bind to DNA and regulate its transcription. For the most part, these physiological changes in gene expression are short lived and are not heritable either from parents to offspring or from parent cells to daughter cells.

Other genetic regulatory mechanisms, known as epigenetic mechanisms, are heritable changes in gene expression and therefore in phenotype that are not dependent on changes in DNA sequence. Three distinct epigenetic mechanisms have been described: DNA methylation, and especially methylation of cytosine residues (recall the role of cytosine methylation in causing C→T mutations); covalent modifications (acetylation, methylation, or phosphorylation) of histones, the major proteins in chromosomes; and expression of the noncoding regulatory RNA molecules we discussed earlier. Epigenetic regulation of gene expression is responsible for the maintenance of distinct specialized cell types in multicellular organisms such as ourselves. Our different cell types have essentially the same genes but have different functions because they exhibit different patterns of gene expression. These patterns of gene expression are stable and are transmitted during replication of our somatic cells. Epigenetic regulation of gene expression is a subject of intense current research. Alterations in epigenetic regulation play a role in cancer and will undoubtedly turn out to be important in other diseases (Tsai and Baylin 2011). Epigenetic mechanisms are also important for X-chromosome inactivation. This is the process by which the expression of most genes on one of the two X chromosomes in the somatic cells of female mammals is inhibited, thereby reducing the effective "dosage" of these genes (Barakat et al. 2010). Many epigenetic marks, as they are called, are removed during the formation of germ cells, but some remain. DNA methylation or other epigenetic changes in germ cells can result in genomic imprinting, the differential expression of genes depending on their parent of origin (Bartolomei and Ferguson-Smith 2011) and may also lead to epigenetic inheritance across generations (Jablonka and Raz 2009).

Epigenetic inheritance is of course dependent on the genes that encode DNA methyltransferases, the enzymes that modify histones, and noncoding regulatory RNA molecules. So in one sense, epigenetic is genetic. Nonetheless, recognition of epigenetic inheritance is an invaluable reminder that heredity involves more than simply the transmission of DNA sequences, or genes, from parents to offspring (Jablonka and Lamb 2005). Reproduction of multicellular organisms like ourselves involves the formation of a fertilized ovum, or zygote. This zygote carries parental DNA sequences but these DNA sequences may be methylated, bound to regulatory RNA sequences, and embedded in chromosomes in ways that result in epigenetic inheritance. In addition, the zygote contains membranes, organelles, and cytoplasmic proteins, which are also crucial for development. Almost all of these other structures come from the oocyte and are encoded by the mother's genes. At birth, newborn infants ingest bacteria from their mothers and thereby inherit their mothers' microbiome. And after they are born, infants and children develop and acquire cultural information in an environment that has been shaped by their parents and others in their communities. All of this additional information contributes to the inheritance associated with human reproduction.

3.9 Population structure and mating patterns

Genotype frequencies in populations are determined not only by the evolutionary processes discussed above but also by patterns of mating within the population. The Hardy-Weinberg equilibrium, proposed independently by the British mathematician G. H. Hardy and the German physician Wilhelm Weinberg, is an idealized model of population genetics that provides a good starting point for thinking about the relationship between allele frequencies and genotype frequencies. This model has many assumptions that are well discussed in textbooks of population genetics (Hartl and Clark 2006). In brief, imagine an autosomal genetic locus with two alleles, A and a, and three genotypes, AA, Aa, and aa. Let the frequencies of A and a be p and q, respectively ($p + q = 1$, because all alleles are either A or a), and assume that the three genotypes have equal fitnesses and so are not being affected by natural selection. Then, in a large, randomly mating population, the frequencies of the three genotypes, AA, Aa, and aa, will be given by p^2, $2pq$, and q^2. For example, if the A and a alleles each have a frequency of 0.5, the AA, Aa, and aa genotypes will have frequencies of 0.25, 0.5, and 0.25, respectively. The model can be expanded to sex-linked genetic loci, to situations in which there are more than two alleles at a locus, and to a consideration of the genotypes formed by alleles at more than one locus.

The Hardy-Weinberg model is important because we can often measure allele frequencies and genotype frequencies in populations. Deviations from the Hardy-Weinberg equilibrium values of genotype frequencies indicate that at least one of the assumptions of the model is being violated. Selection and nonrandom mating are two important reasons why genotype frequencies may deviate from Hardy-Weinberg predictions. A deficiency of heterozygous organisms might be due to selection against this genotype or to assortative mating, a common pattern of nonrandom mating in which organisms tend to mate preferentially with others of the same genotype. An excess of heterozygotes might be due to selection for this genotype (heterozygote advantage) or to the situation in which AA individuals mate preferentially with aa individuals.

3.10 Genetic consequences of human evolutionary history

Homo sapiens is a young species, probably not more than 10 000 generations old. As a result, we are a genetically homogenous species. On average, the genomes of any two people differ in only about 0.1% of their nucleotides. Our ancestors apparently migrated widely throughout Africa before they left that continent or at least before they formed permanent settlements outside of Africa. As human groups migrated south and west from East Africa into other parts of the continent, they became divided into geographically and reproductively isolated populations, which began to differ genetically and phenotypically from one another for reasons we have already discussed—new mutations, natural selection acting on populations in different environments, genetic drift, and direct effects of the environment on developing infants and children. Gene flow among the populations was sufficiently small that the populations remained genetically different—that is, they had different allele or genotype frequencies. Geographic isolation is a major cause of reproductive isolation in humans as it is in most species. But our species is unusual in that it has many reproductively isolated populations whose isolation is based on religion, ethnicity, or social class rather than on geography.

The human population that left Africa and migrated into Asia some 50–100 000 years ago came from East Africa, from around what is now Ethiopia. This was a small population, perhaps numbering in the thousands, and its gene pool contained only a fraction of the genetic diversity that must have already been present in Africa at the time. As a result of this founder effect, non-African populations have less genetic diversity than do African populations. The initial path of human migration out of Africa apparently went first to Western Asia and then along the southern coast of Asia, reaching Australia about 45 000 years ago. Later migrations went from Northwest Asia into Europe and from Northeast Asia into North America. As human groups spread out from Africa to inhabit the rest of the globe, they too formed geographically and reproductively isolated populations that came to differ genetically from one another. Most of the genetic variation in the human species as a whole is variation within populations; only a small fraction is due to variation among populations (Rosenberg et al. 2002). Remember that the genetic differences among populations are differences in allele frequencies. There are occasional "private polymorphisms," new mutations that are restricted to one population. By the time new mutations have spread widely in their population of origin, migrants are likely to have brought them to neighboring populations. Moreover, allele frequencies tend to change gradually as a function of geography (latitude, altitude, etc.), which is what we might expect if reproductive isolation is based in large part on geographic isolation (Cavalli-Sforza et al. 1994). These gradual changes in allele frequencies with geography are known as clines. Our species comprises a nested hierarchy of populations within populations. The largest groups correspond to the major continental divisions but each of these is made up of many smaller, genetically distinct populations.

The dramatic growth of the human population since the agricultural revolution, and especially since the industrial revolution, has led to the production of many new, rare alleles (Keinan and Clark 2012; Reich and Lander 2001). The fate of these alleles is determined more by genetic drift than by selection. Some of these alleles may be associated with disease but their low frequency complicates the identification of these associations. More recently, the rise in international travel and the breakdown of older cultural barriers to intermarriage are leading to an increase in gene flow and thus to a decrease in the genetic differences among human populations. Even if there were no selection, there would be ongoing changes in the genetic makeup of human populations.

3.11 Natural selection in human populations

Although the genetic differences among populations are mainly the result of genetic drift of neutral or near-neutral alleles, the biologically and medically important differences are due to new mutations and to natural selection acting on populations in different environments. Understandably, there has been great interest in identifying genetic loci and their associated phenotypes that have recently been or are currently under selection in human populations. In the nongrowing populations that have characterized humans for most of our history, natural selection would be expected to increase the frequency of genotypes that have the highest survival and fertility rates. Given the high infant and childhood mortality rates that prevailed throughout human history and that continue in many parts of the world today, selection for increased survival, either by increasing resistance to infectious diseases or enabling utilization of new foods, was, and remains, especially important. In developed countries, where almost all newborns survive to reproductive maturity, there is little heritable variation in mortality rates but natural selection may still be acting to increase fertility (Byars et al. 2010).

Several different approaches have been used to study natural selection in human populations. Some studies have correlated phenotypes with the environments in which these phenotypes are thought to be advantageous. Global variations in skin pigmentation are correlated with the strength of ultraviolet (UV) radiation in different environments. Darkly pigmented skin has been selected in populations exposed to high levels of ultraviolet radiation because it protects folic acid from destruction, while light skin has spread in low-UV environments because it promotes the synthesis of vitamin D. Deviations from the correlation of skin pigmentation with UV irradiation can be understood as the result of specific dietary or historical factors (Jablonski and Chaplin 2010).

The availability of genetic sequencing has led to the development of methods to detect what are known as genetic "signals of selection." These signals are based on patterns of genetic variation within and among populations, and on the structure of haplotypes. Recall that as a beneficial allele increases in frequency as a result of selection, it carries with it nearby linked markers, leading to the spread of an extended haplotype block within the population. All of these methods depend upon statistical tests to distinguish the effects of selection from those of genetic drift. These genetic methods have led to the identification of many alleles that have been under recent positive selection. As we might have predicted, these alleles include those that influence skin pigmentation or other adaptations to the physical environment, enable the utilization of novel food sources, or increase resistance to infectious diseases (Hancock and Di Rienzo 2008; Oleksyk et al. 2010).

Natural selection is continuing to act on the human population. Many traits that remain roughly constant, such as birth weight, are maintained constant by a process known as stabilizing selection. Babies with either high or low birth weights have increased rates of infant mortality and so alleles associated with these traits are selected against. As we have discussed, the high infant mortality rates in developing countries must be leading to selection for alleles that are associated with increased survival. The HIV/AIDS epidemic is too recent to have led to measurable changes in allele frequencies but it must be selecting for alleles that confer resistance to this disease. Finally, longitudinal studies such as the Framingham Heart Study are providing direct evidence of natural selection in contemporary human populations (Stearns et al. 2010). Clearly, we remain part of the natural world and continue to be subject to the evolutionary processes that have shaped our history.

4

Cystic fibrosis

4.1 Introduction

In 1909, shortly after Mendel's research became widely known, Sir Archibald Garrod described a group of diseases—alkaptonuria, albinism, cystinuria, and pentosuria—that were transmitted within families and were inherited as recessive Mendelian traits (Garrod 1909). Garrod called these diseases "inborn errors of metabolism." Since that time, thousands of hereditary or genetic diseases, each resulting from mutations in individual genes, have been described. Most of these monogenic disorders are rare, because the alleles that cause them are kept at low frequency by mutation-selection balance. As a group, however, these disorders are responsible for a significant burden of disease. Online Mendelian Inheritance in Man (OMIM), a website maintained by Johns Hopkins University and the National Center for Biotechnology Information, is an invaluable source of information on genetic diseases (OMIM no date).

Genetic diseases seemed initially to resemble the traits studied by Mendel. The disease phenotypes were clearly different from the normal or healthy condition, and they each were apparently caused by one gene with two alleles, normal and mutant. According to the essentialist view of disease that prevailed at that time, the essence of these diseases was the presence of the mutant allele. Just as Mendel ignored individual variations within his groups of plants, physicians did not pay close attention to phenotypic variations among patients with a genetic disease. The etiology and pathogenesis of these diseases seemed to be straightforward: a single mutation resulted in the synthesis of a protein with altered or deficient biological activity, and the change in the activity of this protein in turn led to an altered phenotype or disease. We now recognize, however, that these hereditary diseases are much more complex than was originally thought. We are beginning to understand their genetic and phenotypic heterogeneity, the relationship between genotype and phenotype, and the reasons for the prevalence of some of these mutant alleles in the population. Cystic fibrosis can serve as an exemplar of these diseases.

Cystic fibrosis is often billed as the most common severe autosomal recessive disease affecting Caucasian populations (Welsh et al. 2001). The prevalence of cystic fibrosis in Western European populations and in Caucasian populations in the United States is on the order of 1 in 2500 to 1 in 3500. If we assume that genotype frequencies conform to the Hardy-Weinberg equilibrium, we can estimate that the frequency of mutant alleles in these populations is roughly 1 in 50 to 1 in 60 and the frequency of carriers is about 1 in 25 to 1 in 30. The

Evolution and Medicine. First Edition. Robert L. Perlman © Robert L. Perlman 2013.
Published 2013 by Oxford University Press.

disease is much less common (1 in 100 000 or less) in non-Caucasian populations. There are approximately 4 million births/year in the United States. Slightly less than half of these babies, or roughly 2 million, are Caucasian. Given a prevalence of about 1 in 3000 in this population, there will be about 700 new cases of cystic fibrosis/year in the United States. Other estimates suggest that cystic fibrosis affects about 30 000 people in the United States and perhaps 60 000 worldwide.

Despite its relatively high frequency, cystic fibrosis was not identified as a specific disease until 1938. In that year, Dorothy Andersen, a pathologist at Babies Hospital in New York City, described "cystic fibrosis of the pancreas" as a disease of childhood, comprising pancreatic insufficiency, often associated with intestinal obstruction in the newborn period, fatty stools, and recurrent respiratory infections, and resulting in failure to thrive (Andersen 1938). Children with cystic fibrosis have unusually viscous respiratory and pancreatic secretions. Blockage of airways and of pancreatic ducts by these viscous secretions presumably predisposes affected children to develop respiratory infections and pancreatic insufficiency. Andersen reported that respiratory infections in children with cystic fibrosis were due to *Staphylococcus aureus* and *Streptococcus* species. Later, *Pseudomonas aeruginosa* was recognized as an especially important respiratory pathogen in these children.

When cystic fibrosis was initially described, it was an almost uniformly fatal childhood disease. Affected babies died in infancy of intestinal obstruction or later in childhood from respiratory infections. For a number of reasons, including better nutrition and improved prophylaxis and therapy for respiratory infections, many patients with cystic fibrosis now live to adult life; today, the life expectancy of newborns with cystic fibrosis is close to 40. The increased life expectancy of patients with cystic fibrosis has led to the recognition of another manifestation of the disease. Most men with cystic fibrosis are infertile because they have congenital bilateral absence of the vas deferens.

In 1953, Paul di Sant'Agnese and colleagues noted that children with cystic fibrosis were prone to salt depletion and dehydration, and reported that they had elevated concentrations of sodium chloride in their sweat (di Sant'Agnese et al. 1953). This observation led to the measurement of sweat chloride as a screening test for cystic fibrosis and to the recognition that a defect in epithelial chloride transport was a central feature of the disease (Quinton 1983).

All of these manifestations of cystic fibrosis—viscous pancreatic and respiratory secretions, elevated sweat chloride, failure of development of the vas deferens, and the complications of these defects—result from mutations in a single, obviously pleiotropic, gene. In 1989, the gene whose mutation results in cystic fibrosis was cloned (Riordan et al. 1989). The protein product of this gene is an ATP- and cyclic nucleotide-activated chloride channel, which was named the **C**ystic **F**ibrosis **T**ransmembrane Conductance **R**egulator, or CFTR. The CFTR protein is a member of a large and ancient family of ATP-dependent transport proteins (Jordan et al. 2008). CFTR apparently arose early in vertebrate evolution. In teleost fish, CFTR is important for the physiological adaptation to water of different salinities; it promotes chloride absorption in fresh water and chloride secretion in salt water (Marshall and Singer 2002). These functions foreshadow the role of CFTR in salt balance in humans.

4.2 CFTR, the cystic fibrosis transmembrane conductance regulator

Although many questions about the function of CFTR remain unanswered, the known biological activities of this protein can account in a general way for the pathophysiology of cystic fibrosis. The CFTR gene is expressed in epithelial cells in many tissues and the CFTR protein is localized to the apical membranes of these cells. Because CFTR is expressed in many tissues, it is not surprising that it has multiple functions and that CFTR mutations have multiple phenotypic effects. In sweat glands, the CFTR chloride channel is the primary route of chloride reabsorption. Sweat is initially elaborated as an isotonic secretion with ion concentrations similar to those in plasma. As this fluid traverses along the duct of the sweat gland, sodium and chloride are reabsorbed, so that sweat is secreted as a hypotonic fluid. A decrease in CFTR activity leads to decreased chloride reabsorption and to an increase in the concentrations of chloride and sodium in sweat.

In the intestine, the pancreas, and the lungs, the CFTR chloride channel contributes to chloride secretion, which in turn regulates sodium and fluid secretion. A decrease in CFTR activity results in decreased fluid secretion, increased viscosity of these secretions, and, in some children, blockage of the intestine or pancreatic ducts. CFTR appears to participate in normal intestinal fluid and electrolyte homeostasis. Ingestion of salt leads to secretion of the gastrointestinal hormone guanylin, which activates guanylyl cyclase and increases CFTR activity by a cyclic GMP-dependent pathway (Sindic and Schlatter 2006). The increased CFTR activity, in turn, leads to an increase in intestinal salt and water secretion (Beltowski 2001).

CFTR is expressed in the kidney and presumably plays a role in renal chloride secretion but patients with cystic fibrosis have only minimal deficits in renal function. Other renal chloride channels must be able to compensate for the loss of CFTR. Carriers, heterozygous people who have only one functional CFTR allele, may show minimal defects in sweat production and may have a slightly increased risk of developing respiratory or pancreatic disease but for the most part they are indistinguishable from people without CFTR mutations (Weiss et al. 2005). In terms of allowing people to live healthy lives, the wild type CFTR allele is evidently dominant. It isn't clear why genetic dominance evolved at the CFTR locus.

Although CFTR is described as a chloride channel, it has a number of other biological activities. Either because it transports bicarbonate anions directly or because chloride transport is coupled to bicarbonate/chloride exchange in many tissues, the CFTR channel effectively transports bicarbonate as well as chloride. A decrease in CFTR activity results in decreased bicarbonate secretion and a decrease in the pH of pancreatic and respiratory secretions. In addition, CFTR binds to and regulates the activities of other ion channels; perhaps most importantly, it inhibits an epithelial sodium channel. CFTR also interacts with a number of pathogens. It binds to the lipopolysaccharide in the outer membrane of *P. aeruginosa* and helps to initiate an immune response against this bacterium (Pier 2002), it appears to be the receptor that *Salmonella typhi* uses to enter intestinal epithelial cells (Pier et al. 1998), and it plays a role in the secretory diarrhea caused by *Vibrio cholerae* and enteropathogenic *Escherichia coli* (Kopic and Geibel 2010). The multiple functions of the CFTR protein are another reason why mutations in the CFTR gene have multiple phenotypic consequences.

4.3 Genotypic diversity and phenotypic heterogeneity in cystic fibrosis

Cystic fibrosis was initially assumed to be caused by a single mutant allele. DNA cloning and sequencing methods have led to the characterization of CFTR mutations associated with cystic fibrosis and to the recognition of a previously unsuspected genetic diversity at the CFTR locus. To date, more than 1900 CFTR alleles have been identified (Cystic Fibrosis Mutation Database no date). Of these, more than 20% appear to be due to C→T transitions. Most of these alleles have been found in patients with cystic fibrosis and so are presumably disease-causing but some have been found by screening healthy populations and may be selectively neutral (Pompei et al. 2006). By far the most common disease-causing mutation is the so-called ΔF508 mutation, deletion of a trinucleotide that encodes phenylalanine in position 508 of the mature CFTR protein. The ΔF508 mutation accounts for approximately 70% of mutant CFTR alleles worldwide. Several other mutations are each responsible for 1–2% of the disease-causing alleles, while the vast majority of mutations are extremely rare. Given that the ΔF508 allele represents about 70% of all mutant CFTR alleles and again assuming that genotype frequencies correspond to the Hardy-Weinberg model, only about half of the patients with cystic fibrosis would be expected to be homozygous for this mutation ($0.7^2 = 0.49$). The others are compound heterozygotes with two different mutant alleles or are homozygous for rare CFTR alleles. The distribution of CFTR mutations differs in different populations. In the Ashkenazi Jewish population, for example, only about 30% of disease-causing CFTR alleles are ΔF508. Roughly 60% of these alleles carry a different mutation, the W1282X nonsense mutation, which encodes a polypeptide chain termination signal in place of tryptophan (Trp→Stop).

Many of these mutant CFTR alleles have now been expressed in vitro and their effects on CFTR function have been investigated. Different mutations have been found to affect virtually every aspect of CFTR synthesis and activity. Some, such as the W1282X mutation, prevent the synthesis of any active CFTR; others, such as promoter mutations, decrease the amount of CFTR expression; and still other, missense mutations result in the synthesis of a CFTR protein with altered regulatory properties or chloride channel conductance. The product of the ΔF508 mutation is translated into a full-length protein (missing only one amino acid) but this protein folds abnormally and most of it is degraded within the endoplasmic reticulum. Only small amounts are transported to the plasma membrane, and the protein that does reach the plasma membrane has decreased chloride channel activity and is rapidly removed by endocytosis. Evidently, a single mutation can affect several aspects of the physiology of CFTR. CFTR has multiple domains, regions that are required for the various biological activities of the protein. Some mutations, such as W1282X and ΔF508, result in a loss of all of the biological activities of CFTR. Others, however, may retain significant biological activity or may disrupt some functions of CFTR but leave others intact.

4.4 Relationship between genotype and phenotype

The large number of CFTR mutations and the diverse effects of these mutations on CFTR function help to account for the phenotypic variation among people who carry these mutations. The manifestations of CFTR mutations are roughly correlated with the amount of residual CFTR activity in the affected individuals. The developing vas deferens is most sensitive to loss of CFTR function; reduction of CFTR activity to 10% of normal may result in congenital absence of the vas deferens. In contrast, pancreatic function is relatively resistant to loss of CFTR activity. Symptoms of pancreatic insufficiency typically do not appear unless CFTR activity is reduced to about 1% of normal. Some men with CFTR mutations who showed no signs of cystic fibrosis or other diseases in childhood now seek medical care as adults because of infertility. Other people with mildly deleterious CFTR alleles may also be healthy during childhood but develop pancreatic or lung disease as adolescents or adults.

Although the genetic diversity of CFTR mutations and the multiple functions of the CFTR protein can explain much of the phenotypic diversity among patients with cystic fibrosis, individuals with the same CFTR genotype may not express identical manifestations of the disease. Rudolf Virchow's conception of disease as "life under altered conditions" (Porter 1998, p. 343) provides a fruitful way of thinking about genetic diseases. The manifestations of cystic fibrosis represent life without CFTR. They result from the activities of the other 21 000 genes in the growing embryo, fetus, and child, who is developing in a complex and changing environment. Any genetic or environmental factor that affects the pathways in which CFTR participates may influence the phenotype of people who carry mutant CFTR alleles. Secondary smoke and respiratory pathogens are especially injurious to people with cystic fibrosis. Differences in exposure to these agents are an important reason for differences in the course of cystic fibrosis. Modifier genes, genes at other loci that interact epistatically with CFTR, may also modify the phenotypic consequences of CFTR mutations. Several modifier loci that affect the severity of lung disease in people with cystic fibrosis have been described (Cutting 2010). The gene that encodes transforming growth factor β1, or TGFβ1, appears to be one such modifier. Polymorphisms in the promoter region of the TGFβ1 gene result in different levels of TGFβ1 expression. In several studies, alleles that lead to increased expression of TGFβ1 have been associated with increased severity of lung disease in patients with cystic fibrosis. TGFβ1 participates in the regulation of inflammatory responses but the mechanism by which it affects cystic fibrosis lung disease is not known. The TGFβ1 polymorphism is presumably maintained for reasons unrelated to cystic fibrosis. Moreover, although this epistatic interaction between CFTR and TGFβ1 has been replicated in several studies and appears to be a real phenomenon, it has not been observed in all studies. Epistatic interactions may depend on other genetic or environmental factors that have not yet been identified (Cutting 2010).

4.5 Evolution of mutant CFTR alleles

We can understand the proximate cause of cystic fibrosis, then, in terms of the biological activities of CFTR, the ways in which CFTR mutations lead to alterations in the synthesis or activity of the protein, and the ways that children develop in the presence of these alterations. We need also to understand the ultimate cause of this disease. Why are these mutant CFTR alleles present in the populations and in the frequencies in which they are found? As noted before, the frequency of mutant CFTR alleles in Caucasian populations is about 1 in 50, or 0.02. The ΔF508 mutation, which accounts for approximately 70% of the mutations in Caucasians, thus has a frequency of about 0.014. Because almost all of the ΔF508 mutations occur on a single haplotype background, all of the ΔF508 alleles present today are most likely descended from a single mutational event. The age of the last common ancestor of the ΔF508 allele remains controversial. The geographic distribution of the ΔF508 allele suggests that it increased in frequency some time after the ancestors of today's European populations became isolated from other populations. The best current estimate is that this allele began to spread at least 600 generations or 15 000 years ago (Wiuf 2001). The geographic site of origin of the ΔF508 allele is also uncertain. It may have arisen in a population living somewhere in Asia, who then migrated into Europe before the allele began to spread (Saleheen and Frossard 2008).

The frequency of the ΔF508 allele in Caucasian populations is much too high to be the result of mutation-selection balance and also seems too high to have been caused by genetic drift. Most cystic fibrosis researchers assume that the ΔF508 allele spread because of heterozygote advantage; in other words, that carriers of this mutation have (or had) increased fitness (Romeo et al. 1989). Many hypotheses to account for this presumed heterozygote advantage, including resistance to a variety of pathogens, reduced rates of asthma, and increased fertility, have been proposed (Poolman and Galvani 2007; Welsh et al. 2001). At present, the most attractive hypothesis is that the ΔF508 allele spread in European populations because carriers of this allele had increased resistance to *Mycobacterium tuberculosis* (Poolman and Galvani 2007). Tuberculosis has been a globally endemic disease since the time of the agricultural revolution and became a major cause of death in Europe in the sixteenth or seventeenth century. Tuberculosis, bubonic plague, and smallpox are probably the only diseases that could have provided the selection pressure necessary to increase the frequency of the ΔF508 allele in European populations to its current levels. CFTR regulates the composition of respiratory secretions, so a relationship between CFTR mutations and *M. tuberculosis* infections is plausible. In contrast, there is no obvious physiological connection between CFTR and either plague or smallpox. The clinical or epidemiological evidence that people who are heterozygous for mutant CFTR alleles are in fact resistant to tuberculosis is weak but anecdotal reports support this hypothesis (Stiehm 2006). Fortunately, economic development is decreasing the prevalence of tuberculosis and most patients who do get the disease can be successfully treated with antibiotics. We may never know the reasons for the spread of the ΔF508 allele. Whatever these reasons may have been, they are no longer operative, and ΔF508 is now a deleterious allele. It remains prevalent today because only recently have its deleterious effects begun to outweigh its beneficial effects and it hasn't yet been reduced to low frequency by natural selection.

The ancestral CFTR gene encodes methionine in position 470 of the CFTR protein. Another CFTR mutation, M470V, encodes valine instead of methionine at this position. The M470V allele is rare in African populations but has spread to high frequency in Asian and European groups. In some populations, it is now more common than the ancestral allele. The protein encoded by the M470V allele has altered chloride channel properties but does not cause cystic fibrosis. The rapid spread of this allele, together with the presence of an extended haplotype block around it, strongly suggests that the high frequency of this allele is due to selection (or hitchhiking) rather than drift (Pompei et al. 2006). The M470V allele has been associated with an increase in male fertility in some populations (Kosova et al. 2010). Increased fertility could explain why this allele has increased in frequency. In contrast, the ΔF508 allele does not appear to increase fertility (Jorde and Lathrop 1988). Given the multiple biological activities of CFTR, different alleles may well affect fitness in different ways. The ΔF508 and the M470V mutations may have spread for different reasons.

While it is attractive to think that the ΔF508 allele, and perhaps several of the other more common CFTR mutant alleles, increased to their present frequencies as a result of natural selection, the myriad of rare CFTR alleles can best be understood as a product of human demographic history. Although the presence of a multitude of rare CFTR alleles was initially unexpected, in hindsight it is just what we should have anticipated. These rare CFTR alleles are younger than the ΔF508 mutation. Most of them probably arose during the dramatic increase in the human population that has occurred over the last several hundred years. Several CFTR alleles are present at high frequency in French Canadian (Quebecois) populations because of founder effects (De Braekeleer et al. 1996) and several others appear to be de novo mutations in affected patients. As we discussed in Chapter 3, expanding populations accumulate large numbers of mutant alleles, regardless of their effect on fitness (Keinan and Clark 2012; Reich and Lander 2001). Even if some of these alleles increase resistance to tuberculosis or increase fitness for some other reason, their current low frequencies are better understood as resulting from genetic drift rather than from natural selection. It may just be a historical accident that the ΔF508 allele was present in European populations at the time tuberculosis became a major cause of death, while these other alleles didn't arise until later.

As the example of cystic fibrosis illustrates, "simple" genetic diseases have turned out not to be simple at all (Scriver and Waters 1999). As in cystic fibrosis, there are often many, many alleles at the disease gene locus, most or all of which are new and rare (Reich and Lander 2001). Other genetic diseases, too, show a wide range of disease severity and, because of environmental factors or epistatic interactions among genes, the correlation between genotype and phenotype is often poor. Identification and study of the underlying genes has made prenatal diagnosis possible and so has improved genetic counseling, and has led to important insights into the pathogenesis of these diseases. Improved medical care has enabled patients with cystic fibrosis and many other genetic diseases to live longer and healthier lives. While these diseases may eventually be treated by gene therapy or by nongenetic treatments that correct or ameliorate the genetic deficits, such therapies have yet to be developed.

5

Life history tradeoffs and the evolutionary biology of aging

5.1 Introduction

We commonly talk about our lives in terms of the life course or life cycle. Although each of us goes through a life cycle only once, there is a continuity of life cycles from generation to generation. Some biologists view life cycles as the fundamental units of biological organization and think about evolution as the evolution of life cycles (Bonner 1993; Sterelny and Griffiths 1999). Evolutionary life history theory provides a framework for understanding how natural selection has shaped life cycles in ways that optimize reproductive fitness (Charnov 1993; Stearns 1992). Life history theory provides a coherent view of the entire human life cycle and is especially important for medicine because it underlies the evolutionary theory of aging.

Organisms require nutrients and other resources—oxygen, water, heat, shelter, other organisms, etc.—to survive and reproduce. They can acquire these resources directly from their environments or indirectly from others, most often their parents, and they have to allocate these resources, as well as their time, among a handful of tasks, including growth and development, reproduction, prevention and repair of bodily damage, external work, and storage. Prevention of bodily damage includes responding to and countering threats and stresses. And reproduction does not simply mean the energetic cost of having babies. It refers to all of the costs of producing and maintaining secondary sexual characteristics and finding mating partners, as well as bearing, raising, and providing resources for their offspring. In demographic terms, organisms use resources to increase survival and to increase fertility. Life history theory is a theory of the ways in which natural selection has shaped the physiological and behavioral mechanisms that regulate the acquisition and allocation of these resources over the life cycle.

The human life cycle can be divided into several more or less distinct stages—prenatal, infant (nursing), post-weaning dependent child, puberty and adolescence, sexual and reproductive maturity, and post-reproductive life. These stages are distinguished by differences in the ways we acquire and use nutrients and other resources. Natural selection has adjusted the timing of these stages as well as our activities in them. It has modulated the duration of gestation and of nursing, our overall growth rates and the growth of individual organs during childhood, the timing of puberty and the age at which we reach reproductive maturity, how

Evolution and Medicine. First Edition. Robert L. Perlman © Robert L. Perlman 2013.
Published 2013 by Oxford University Press.

frequently we reproduce and how many children we have each time we do so, the age at which we cease reproduction, and the length of our post-reproductive lives (Stearns et al. 2008). Hormones mediate many of these life history "decisions" and so they play a central role in life history theory.

Each of these life history decisions involves tradeoffs or compromises. With respect to the duration of gestation, for example, there are advantages for fetuses to remain in utero, since the uterus provides a protected environment in which they can grow and develop. As fetuses grow larger, however, they become a bigger nutritional burden on the mother and suffer an increased risk of complications during labor and delivery. The timing of labor and delivery is apparently the optimal compromise between these two competing demands (Ellison 2001). The age at which we reach puberty and sexual maturity also involves trade-offs. We need time to develop physically, psychologically, and socially to the point that we can bear and care for children. On the other hand, we need to become sexually mature and have the opportunity to reproduce when we are still young enough that we can have and raise our children before we're likely to die. Yet again, there is a tradeoff between fertility and the survival of offspring that we discussed in Chapter 2, because the more children we have, the fewer resources we can allocate to each and the greater is their risk of dying before they reach reproductive maturity.

Presumably because they are regulated by pleiotropic genes, many life history traits are correlated among primate or mammalian species. For example, the duration of childhood or juvenility among primates is correlated with their life expectancy (Leigh 2001). There may be tradeoffs among suites of traits that evolve together. Moreover, as we shall discuss later, our life history strategies are not fixed but can change in response to environmental signals we receive during development. The range of life histories or other phenotypes that can be produced by organisms of a single genotype developing in a range of environments is known as their norm of reaction (Gilbert 2001). Natural selection does not act simply to optimize fitness in one specific environment. It modulates the norm of reaction to optimize fitness across the diversity of environments that a species has encountered during its evolutionary history. Our developmental flexibility and our ability to survive and reproduce in different environments may limit our adaptation to any one specific environment.

The human life cycle resembles the life cycles of other great apes, especially chimpanzees, but has a number of special features that presumably arose during the past 6 million years or so, since the time the hominin (human) lineage diverged from the lineage leading to chimpanzees (Robson et al. 2006). We have a slightly longer gestation period (9 vs 8 months) and lower rates of twinning than do chimpanzees. Women in many traditional societies nurse their babies for 2½–3 years, whereas chimpanzee mothers nurse for 4–5 years. Juvenile chimpanzees are able to acquire much of their own food by the time they are weaned and typically leave their mothers 2–3 years after weaning. We have a much longer period of childhood or juvenility, when children are no longer nursing but are still dependent on parents or others for food and survival. Because of this long period of childhood dependence, women typically have to care for several dependent children at the same time. Women have their first babies at a considerably older age than do chimpanzee females (19–20 vs 13–14 years). Because of earlier weaning, women in natural fertility societies, i.e. in the absence of contraceptives or family planning, then have babies at shorter intervals than do chimpanzees (3–4 vs 5–6 years). Finally, although both women and chimpanzee mothers can have offspring until their early forties, we have much longer post-reproductive lives: chimpanzee females have life spans of 50–55 years while women have life spans of 80 or longer.

The special features of the human life cycle must have been shaped by the environments in which ancestral humans evolved, including among other factors the composition of ancestral human diets (Kaplan et al. 2003). Chimpanzee diets are rich in foods such as fruits and piths (the soft, fleshy material in the stems and branches of plants), which are relatively easy to obtain. Although we also eat fruits, they are a smaller component of our diets. We have evolved to eat a variety of energy-rich but difficult to acquire foods, such as nuts, seeds, roots, and meat. It takes a long time for children to acquire the skills needed to collect enough of these foods to be nutritionally self-sufficient. After they acquire these skills, however, adults can gather or hunt much more than they themselves need. Our long period of childhood dependence is made possible by the contributions of adults to the nutritional, educational, and social support of children.

There are controversies over how long it takes children to acquire the skills they need to become self-sufficient and over the relative roles of fathers, grandmothers, and other adults in helping to provision dependent children (Hawkes 2004). These aspects of child development and childcare may well differ in different societies. Despite these unresolved questions, this ecological perspective provides a plausible and attractive way of understanding the evolution of the human life cycle. Evolution of a prolonged childhood, which would have given children the opportunity to gain the knowledge and skills they needed to become self-sufficient, would have been accompanied by evolution of a longer life span, which gave adults more time to provision their dependent children. Long before the recent advances in medicine and public health that have extended our life expectancy, many of our ancestors who survived infancy and childhood went on to have long lives. Even in a population with a life expectancy of only 20 years, about a quarter of the population lives until age 50 (Coale 1974).

5.2 The causes of death change through the life cycle

Different stages of the life cycle are characterized by different patterns of disease and death. Recall that the major causes of death in infancy and childhood include congenital malformations and chromosomal abnormalities, genetic diseases, developmental defects, infectious diseases, and nutritional deficiencies. The high mortality of adolescent and young sexually mature men is due in large part to accidents, violence, and suicide. Deaths due to cardiovascular diseases and most cancers increase from about age ten onward but these diseases don't become the leading causes of death until around age 40, near the end of our period of reproductive maturity.

In addition to cardiovascular diseases and cancer, other noncommunicable diseases, including diabetes, chronic obstructive pulmonary disease, and neurodegenerative diseases, are also major causes of disability and death in the United States and other economically developed countries. As developing countries go through the demographic and epidemiologic transitions we discussed in Chapter 2 and more people survive to reach the post-reproductive stage of the life cycle, these diseases of aging, which we shall refer to later as degenerative and man-made diseases, are bound to become even more important components of the global burden of disease (Omran 1971). These various classes of disease appear to have little in common. They affect different organs, they cause different kinds of pathological changes, and they produce different symptoms and disabilities in people who suffer from them. Understandably, different medical specialties have developed to care for patients with these various diseases and different advocacy groups have arisen to provide support for them. All of these clinically distinct classes of disease do, however, share one important epidemiologic characteristic: the incidence of and mortality from each of them increase with age. Life history theory provides a foundation for understanding many features of these and other diseases of aging.

5.3 What is aging?

Aging, or senescence, may be defined as a progressive, generalized impairment of function, resulting in a loss of adaptive responses to stress, an increasing probability of death, and, in many species, a decline in fertility (Kirkwood 1996). Although this definition sounds grim, we should remember that, at least in economically developed countries, aging is often accompanied by fewer responsibilities and increased leisure time, and that many older people continue to lead meaningful, productive, and rewarding lives even as they show some of the physiological signs of aging. Indeed, surveys suggest that older people are on average happier than younger ones. Aging itself is not a disease; it is a normal and expected—and hoped for—part of the life cycle. As the saying goes, growing old is a privilege. Nonetheless, as we grow older and enter the post-reproductive stage of the life cycle, we do have an increasing risk of developing and dying from a wide variety of seemingly unrelated diseases.

Why do organisms age? Aging did not evolve because aging and death make room for new organisms and so are good for the group or the species. This type of group selection is not a viable evolutionary process, because it would be undermined by selection for traits that increased individual survival and reproduction (Williams 1957). Moreover, aging did not evolve because the entropy, or disorder, of living organisms must inexorably increase according to the dictates of the second law of thermodynamics. In physical terms, living organisms are open systems that exchange matter and energy with their environments. As long as nutrients and energy are available, there is no thermodynamic reason why organisms could not survive and reproduce indefinitely. Nonetheless, if they live long enough, organisms of most species, even some bacteria, show evidence of aging. A number of invertebrate species do not show increases in age-specific mortality rates and so do not appear to age in nature (Rose 1991). We don't understand why these organisms don't age. For the moment, their lack of aging can only be seen as an intriguing curiosity. By and large, it seems that living organisms have evolved life histories that cause them to age.

Understanding the evolutionary biology of aging begins with a recognition that, as organisms go about their lives and interact with their environment, they expose themselves to risks of death from what biologists since Aristotle have called external or extrinsic causes (Aristotle c.350 BCE). External causes of death, which include natural disasters, starvation, and predation, are causes that cannot be eliminated by natural selection or by changes in life history strategies. Unpredictable natural disasters such as tsunamis and earthquakes kill organisms of all genotypes indiscriminately. The organisms that survive these disasters do so because of good luck, not because they have disaster-resistance genes. There may well be genetic variation in risk taking but it's hard to imagine that earthquakes select for risk-avoidance genes. Likewise, organisms have minimal nutritional requirements. Natural selection can enhance the efficiency of metabolism and can produce adaptations such as hibernation or dormancy that reduce metabolic rate when food is unavailable but it can't protect organisms from prolonged periods of drought or famine except perhaps by causing the production of inactive spores. And there are limits to the ability of natural selection to reduce susceptibility to predation within a population, because there are relatively few traits (speed and endurance, primarily) that might affect vulnerability to predation, and because predators coevolve with their prey. Adaptations that reduce predation, such as flying or burrowing, may involve such large changes in behavior or ecology that they lead to the formation of new species.

No matter what genotype they have, then, organisms will die at some rate from natural disasters, starvation, predation, or other extrinsic causes. As a cohort of organisms grows older, it will inevitably become smaller. In nature, most organisms die of these extrinsic causes before they have had a chance to age. Because organisms typically experience high mortality rates early in life, the age structure, or population pyramid, of most populations is heavily weighted toward younger organisms. Even though organisms of most species will show signs of aging if they live long enough, few have the privilege of growing old. And since relatively few organisms in nature experience aging, the characteristics of aging are unlikely to be the direct result of natural selection. Instead, they are almost certainly the incidental byproducts of selection for other traits, specifically selection for traits that increase reproductive fitness (Kirkwood and Austad 2000).

The rates at which populations die from extrinsic causes play an important role in shaping their life histories. Recall the tradeoffs between growth and development, reproduction, and bodily repair. Because of these tradeoffs, natural selection will cause populations that experience high mortality rates to evolve to reproduce early, such that members of these populations will reach reproductive maturity at ages when they still have a relatively high probability of survival. In contrast, populations with lower levels of mortality will, in general, evolve such that their members can allocate more resources to growth and development and delay reproduction (Stearns 1992).

5.4 The life history theory of aging

The life history theory of aging is based on the idea that the ability of natural selection to prevent death decreases with age. The evolutionary biologist William Hamilton was the first person to point out clearly that the force of natural selection acting on age-specific mortality rates is proportional to the expected future contributions of this age group to their reproductive fitness (Hamilton 1966). Imagine an allele that causes death at a given age. If the allele causes death in infancy or childhood, before the onset of reproductive maturity, it will reduce fitness to zero and will be strongly selected against. As we saw in Chapter 3, the process of mutation-selection balance keeps such alleles at very low frequency. On the other hand, if the allele causes death after the end of our reproductive period, it will have no effect on fitness and will not be selected against. Between these extremes, there is a continuous decline in the ability of natural selection to prevent death at a given age. In brief, organisms age and die because of this decline in the ability of natural selection to prevent death.

Recall that the age-specific fertility rate for women declines to very low levels by age 45 or so. But our reproductive period does not end with the birth of our last child. Successful reproduction requires child rearing and perhaps even grandchild rearing. This extended period during which survival contributes to reproductive fitness has presumably slowed the decline in the force of natural selection and has contributed to the evolution of a significant life expectancy after menopause (Williams 1957). Nonetheless, there must be some age by which we have borne and raised our children and have helped ensure the survival of our grandchildren, and beyond which our survival has no evolutionary consequences—that is, beyond which death does not decrease fitness. At this age, the ability of natural selection to decrease mortality rates declines to zero.

The force of natural selection in preventing decreases in fertility also declines with age (Hamilton 1966). Changes in fertility are more difficult to analyze than are changes in mortality. Recall the tradeoff between fertility and offspring survival, between the quantity and quality of offspring. A decrease in fertility at one age may conserve resources that can be used to increase maternal survival and future reproduction. Moreover, as we discussed with regard to the demographic transition, fertility rates may be influenced by cultural norms as well as by natural selection. We shall focus on those aspects of aging that lead to increased mortality rather than those that result in decreased fertility.

If a population experiences high extrinsic mortality rates and evolves to have early reproductive maturity, the force of natural selection will decline rapidly with age. The success of a population in decreasing or preventing extrinsic causes of death will slow the decline in natural selection with age. Thus, the ecological interactions of organisms with their environments play a major role in shaping the life histories of their species. This evolutionary/ecological view of aging helps to explain many features of the comparative biology of aging. Among mammals, size and life span are closely correlated—large mammals live longer than small ones. We can understand this relationship because, in general, large animals are less subject to predation and so have reduced extrinsic mortality. Elephants, which are almost immune to predation, have exceptionally long life spans. Bats live longer than do other comparably sized mammals. Again, this makes sense, because bats can escape predation by flying and perching in inaccessible places. Perhaps for the same reason, birds live longer than do comparably sized mammals. And animals with extremely long life spans, such as some sea turtles and tortoises, have protective shells that greatly reduce their mortality from predation.

Among our close evolutionary relatives, humans have much longer life spans than do chimpanzees or gorillas. Recall that the life span of chimpanzees is in the fifties (Rose and Mueller 1998). Our life span has almost doubled during the roughly 6 million years since the hominin and chimpanzee lineages diverged. An attractive hypothesis to explain our increased longevity is that, as our ancestors evolved larger brains, developed the ability to work cooperatively in groups, and acquired the capacity for the transmission of cultural information, we became better able to hunt, to find food, and to avoid or kill predators. These skills decreased extrinsic mortality, which led to the evolution of delayed reproduction, which in turn slowed the decline in the force of natural selection with age. Our slow rate of aging and our prolonged childhood evolved in concert with our evolutionary ancestors' increasing success in warding off causes of death to which *their* ancestors would have succumbed. Again, our health and long lives are not the results of direct selection for longevity but are byproducts of selection for reproductive fitness. Our ancestors evolved to stay alive and healthy long enough to produce and raise their own children, and perhaps to help raise their grandchildren. What happened to them after that was of little evolutionary consequence.

5.5 Genetic causes of aging

There are several ways in which the waning power of natural selection might lead to aging. One early idea, proposed by the immunologist Peter Medawar, is the hypothesis of mutation accumulation (Medawar 1952). Medawar recognized that, as we have just discussed, mutations that increased the probability of death in childhood would be strongly selected against, whereas mutations that increased the probability of death in older people would not be. Deleterious mutations that act at older ages may simply accumulate in a population by genetic drift. The mutation accumulation hypothesis can account for the existence of some genetic diseases, such as Huntington's disease, which typically strikes people in their forties or fifties, but late-acting alleles seem unlikely to play a major role in aging. Most genes act early in development; few if any have no phenotypic effects until late in life.

Another hypothesis, that of antagonistic pleiotropy, was advanced by the evolutionary biologist George Williams (Williams 1957). Williams proposed that some alleles which had beneficial effects in childhood were detrimental in old age. Such alleles would spread in a population if their beneficial effects early in life outweighed their later harmful consequences. It is easy to imagine the existence of alleles with these properties. Williams' example was a hypothetical allele that increased bone calcification during development but led to increased arterial calcification later in life. Most genes are pleiotropic and their net effect on fitness must involve a tradeoff among all of their beneficial and deleterious effects.

More recently, the antagonistic pleiotropy hypothesis of aging has been formulated explicitly in terms of evolutionary life history theory. Tom Kirkwood, who is one of the architects of this view of aging, has dubbed it the disposable soma hypothesis (Kirkwood and Austad 2000). In today's idiom, it might be thought of as the recyclable soma hypothesis (Perlman 2008). Again, we have evolved mechanisms that allocate our finite resources of energy and time in ways that enhance our reproductive fitness. The utilization of these resources necessarily involves tradeoffs between survival and fertility. Any genes that influence the allocation of resources must have pleiotropic effects: energy that is used for reproduction cannot also be used to prevent or repair bodily damage. Because resources are finite and because some must be allocated for other purposes, bodily maintenance and repair is in general not perfect. As a result, cells die, organ function declines, the ability to respond to stress is decreased, and the probability of death increases—in other words, we age. We don't have genes "for" aging. Instead, we have genes that regulate the allocation of energy to bodily repair and other purposes in ways that enhance fitness. Aging is the byproduct of this resource allocation. Natural selection has optimized our ability to transmit our genes to our children and grandchildren but once we have completed that task, our bodies are disposable and are recycled. Although he was writing with a different purpose, T. S. Eliot described the life history view of aging memorably: "And so each venture is a new beginning . . . With shabby equipment always deteriorating" (Eliot 1943).

5.6 Proximate causes of aging

From an evolutionary point of view, then, aging is a consequence of the evolved allocation of resources in ways that maximize our reproductive fitness even at the expense of somatic repair. What are the proximate mechanisms of somatic breakdown that require maintenance and repair? Our bodies are constantly undergoing somatic damage. Ions leak across cell membranes and have to be pumped in and out of cells; proteins become denatured and nonfunctional and must be degraded and replaced; DNA undergoes chemical damage (e.g. mutation) and has to be repaired; lipids suffer oxidative damage, etc. Many cell populations—epithelial cells such as those in the skin and gastrointestinal tract, red blood cells, and lymphocytes, among others—have limited life spans. These cells die and must be replaced. Finally, our bodies lose heat to the environment, and we need energy to overcome this heat loss and maintain our body temperature. All of these repair and maintenance processes require energy that might otherwise be used for reproduction or other purposes.

Much of the damage to somatic tissues comes from oxygen and glucose, two of the major substrates of metabolism. Both oxygen and glucose are chemically reactive molecules, which is presumably why they came to play essential roles in most living organisms. Nonetheless, their chemical reactivity can result in somatic damage. The spontaneous conversion of oxygen to superoxide radicals and other so-called "reactive oxygen species" leads to the oxidation of many cellular constituents. The oxidative deamination of cytosine residues in DNA that we discussed earlier is just one of several kinds of oxidative DNA damage that will result in somatic mutations and the loss of normal cellular function if this damage is not repaired before the DNA is replicated (Fraga et al. 1990). Reactive oxygen species can also oxidize and damage proteins and lipids, which in turn disrupts the structure and function of cell membranes. Energy must be used to remove and replace these oxidized lipids. Glucose reacts with free amino groups in proteins (particularly the N-terminal amino groups), resulting in glycation of these proteins. Glycation of collagen and elastin leads to covalent cross-linking and to the loss of elasticity seen in the skin of older people, and glycation of lens proteins may lead to cataracts. But these are just the most obvious signs of the kinds of chemical damage that presumably impairs the function of other, more vital, proteins. Mitochondrial damage, especially deletions of mitochondrial DNA (Wallace 2010), and shortening of telomeres, the ends of chromosomes (Shammas 2011), can also lead to loss of cellular function.

In summary, our metabolic processes, which provide the energy we need for growth and development, for reproduction, and for work, are accompanied by damage to cells and organs, and some of the energy derived from metabolism must be used to minimize or repair this damage. Imperfect prevention and repair leads to the accumulation of bodily damage, which in turn leads to aging. Understanding aging as a consequence of our evolved life history strategy is consistent with and complements mechanistic theories of the aging process, which focus on the proximate mechanisms of the loss of bodily functions. An evolutionary understanding explains why we age and why organisms of different species age at different rates, and it may suggest interventions to decrease the rate of aging and postpone the onset of diseases of aging.

Given that aging results from utilization of limited nutritional resources for reproduction and other purposes rather than for somatic repair, it may seem paradoxical that calorie restriction increases longevity in many species (Bishop and Guarente 2007). One answer to this paradox appears to be that glucose and oxygen are responsible for much of the wear and tear of living. Calorie restriction may decrease the production of reactive oxygen species and so may decrease the rate of somatic damage, and may also divert energy from reproduction to bodily maintenance. The sex hormones testosterone and, to a lesser extent, estradiol decrease immune function and redirect resources to reproduction. Given this tradeoff between survival and fertility, it is not surprising that, in many species, including humans, castration increases longevity (Hamilton and Mestler 1969; Min et al. 2012). Less drastically, reproduction (number of children) appears to be negatively correlated with longevity in some human populations (Doblhammer and Oeppen 2003; Westendorp and Kirkwood 1998).

5.7 Somatic repair and the depletion of physiological capital

As life history considerations might lead us to expect, we have evolved a wealth of biochemical mechanisms to minimize and repair somatic damage. Natural selection has adjusted both the rate of somatic damage and the activity of our repair mechanisms. Our basal metabolic rate—our bodies' minimum caloric needs—is governed largely by the energy we devote to somatic repair. A significant amount of our basal or resting metabolism is used to pump ions across cell membranes and maintain the ionic composition of cells within physiological limits. Another substantial amount of energy is needed for protein synthesis and degradation. And part of our basal metabolism provides energy needed by the respiratory and circulatory systems to deliver oxygen and other nutrients to our tissues.

Antioxidant enzymes comprise an important group of bodily defense and repair mechanisms. Enzymes such as superoxide dismutase degrade reactive oxygen species and limit the oxidative damage these molecules would otherwise cause. The levels of superoxide dismutase and other antioxidants in mammalian tissues are correlated with longevity (Cutler 1991). Although these correlations do not prove a causal relationship, they are consistent with our evolutionary understanding of aging. We also have enzymes that reverse oxidative damage to proteins and nucleic acids. Given the central role of DNA in cell function, it is not surprising that we have a myriad of mechanisms that maintain the integrity of DNA and minimize the production of mutations. The activity of DNA excision-repair enzymes is also correlated with longevity (Hart and Setlow 1974). Again, these mechanisms have not evolved to be perfect. Presumably, they have evolved to slow the rate of bodily damage to a level that optimizes our reproductive fitness.

Rates of aging are determined not only by the rates of bodily damage and repair, but also by the physiological capacities, or physiological reserves, that we accumulate during fetal and early postnatal development. For example, we are born with stem cells in many tissues. These stem cells can differentiate throughout life to replace cells that are damaged and destroyed by the mechanisms we have described. Eventually, however, our supply of functional stem cells becomes depleted and our ability to replace dead or damaged cells is decreased. Likewise, we are born with many more nephrons in our kidneys than we need to maintain physiological homeostasis. Indeed, we are born with two kidneys but we need only one to remain healthy. During life, nephrons become damaged and nonfunctional, and kidney function declines, but we do not develop signs of renal insufficiency until kidney function has decreased to perhaps 20% of its initial level.

An economic metaphor may help to clarify our understanding of aging. The endowment of stem cells, nephrons, and other resources that fetuses and infants acquire during development comprises what the economist Robert Fogel has called physiological capital (Fogel 2003). Physiological capital provides a reserve of physiological capacity that buffers us against the damage that occurs during life. Natural selection has provided us with sufficient physiological capital to stay alive and reproduce successfully in the face of the expected demands for this capital. Eventually, as our physiological capital becomes depleted, we get sick and die. The rate at which we age is determined by the amount of physiological capital we accumulate early in development and the rate at which this capital is depleted as a result of the physiological damage we suffer later in life.

From a public health perspective, one of the most common and debilitating deficits in the elderly is a loss of balance and an increased risk of falling. Falls are not only physically damaging but fear of falling may cause older people to become less active and more socially isolated. Our sense of balance depends on the proper functioning of the otoconial organs, the utricle and saccule, in our inner ear. Hair cells in these organs have cilia that are embedded in an extracellular matrix made up of otoconia, macromolecular complexes containing calcium carbonate crystals bound to glycoproteins. Otoconia are gravity receptors. They are heavier than extracellular fluids. When they move in response to gravity they displace the cilia, which leads to changes in the neuronal output of the otoconial organs. Our utricles and saccules contain several hundred thousand otoconia, which are laid down during fetal development before these organs are functional. Over time, calcium is leached from the otoconia into the surrounding extracellular fluid and the otoconia fall apart. Because these damaged and nonfunctional otoconia can't be replaced, we lose our sense of balance. Changes in otoconial function during the life cycle exemplify the concept that the impairments of aging result from the balance between the accumulation and depletion of physiological capital. Unfortunately, we don't yet know how to modulate either the accumulation of otoconia during development or their rate of loss during life (Mason 2011).

Different people manifest the effects of aging differently. Some develop cardiovascular failure, some show signs of respiratory insufficiency, some suffer cognitive loss, and some acquire other deficits or disabilities. These differences are due in part to environmental factors, such as occupational hazards, nutrition, or behavior, that may increase damage to one specific organ or organ system. In addition, genetic risk factors increase the susceptibility of different people to different diseases. These genes may affect the accumulation or depletion of specific types of physiological capital. Somatic mutations may play an important role in the development of age-related diseases. For the population as a whole, however, aging seems to involve a more or less simultaneous loss of function of many organs and organ systems. This simultaneous wearing out of many physiological systems can be understood in terms of what Richard Dawkins has called the principle of animal design (Dawkins 1995). Natural selection is unlikely to produce kidneys that can sustain us for 150 years if our hearts are likely to stop functioning after age 80. Again, there are tradeoffs in the use of limited resources; energy that is used to develop physiological reserves that will never be needed can better be used for other purposes. Natural selection presumably adjusts the activities of antioxidant defenses, DNA repair enzymes, etc., so that different types of unrepaired somatic damage all contribute to aging and disease. An evolutionary view of aging helps us understand why older people tend to suffer from multiple ailments, or comorbidities. It also helps us appreciate that preventing or curing the major age-related diseases will not lead to immortality. People who don't die from one disease of aging are likely to die of another before very long. Finally, it helps us realize why the multiple parallel pathways that contribute to somatic breakdown—oxidative damage, mitochondrial dysfunction, etc.—have made it difficult to identify the relative importance of specific causal mechanisms in aging.

In short, evolutionary life history theory provides a firm foundation from which to understand the important diseases of aging. An appreciation that aging results from our evolved life history strategies, from the balance between the accumulation and depletion of physiological capital, provides a framework around which to design interventions to slow the aging process and postpone the onset and progression of diseases of aging.

5.8 Plasticity in rates of aging

Although our aging may be inevitable, the human life span and the time course of aging are not fixed. During the past century, the life expectancy of 50-year-old people in the United States has increased by 10 years, from 21 years in 1900 to 31 years today. There are several reasons why life expectancy, even of older people, is continuing to increase. Presumably because of better maternal health and nutrition, infants are born with more physiological capital and so have healthier or more resilient cells, tissues, and organs. Better nutrition and fewer serious infectious diseases in infancy and childhood preserve our physiological capital and slow its decline. Thus, people are entering adult life better able to respond to or repair the insults that lead to the diseases of aging. Environmental hazards such as smoking are major causes of somatic damage. The decline in smoking over the last 50 years has contributed significantly to our increased life expectancy (Peto et al. 2000; Stewart et al. 2009). Older adults today have fewer chronic conditions and get these diseases later in life than did adults a century or even a generation ago (Fogel 2003). This postponement of the onset of diseases of aging is probably due both to greater reserves at birth and to slower declines during life.

5.9 Developmental origins of health and disease

Plasticity in rates of aging is part of a broader plasticity in our life history strategies and adult phenotypes. Recall that natural selection shapes organisms' developmental processes in ways that optimize their norm of reaction, their fitness over a range of environments. Developing organisms receive signals or cues about their environments. Zygotes receive epigenetic marks on the genomes they acquire from their parents at the time of fertilization, and fetuses receive nutritional and endocrine signals from their mothers throughout their development. These signals presumably convey information about the mother's health and nutritional status, as well as about other aspects of the mother's environment. They may cause changes in development that prepare the fetus for the environment it will experience after it is born or possibly for the environment it is likely to encounter later in life (Bateson 2001).

Just as extrinsic mortality rates have shaped the life histories of species over evolutionary time, anticipated mortality rates may modulate individual life history strategies during development. Female fetuses and infants who receive information suggesting that they will grow up in a harsh or dangerous environment, where the expectation of a healthy life is short, develop in ways that lead to earlier menarche and tend to have their first children at earlier ages (Chisholm 1999; Nettle 2011; Sloboda et al. 2009). The environmental signals that lead to early puberty and early reproduction may be nutritional, endocrine, or psychosocial, and they may act in utero or in childhood. The physiological changes associated with early menarche presumably involve epigenetic changes in gene expression that lead to early activation of the hypothalamic-pituitary-ovary axis. Increased investment in early reproduction is accompanied by decreased investment in physiological capital and somatic maintenance, and so results in an increased risk, or earlier onset, of diseases of aging. Even though this life history strategy of early puberty and early reproduction is associated with an increased risk of adult disease, it is probably best understood as an adaptive response that evolved because it optimized reproductive fitness in stressful or dangerous environments, where it may have been risky to postpone reproduction. This phenomenon appears to be an example of what might be called epigenetic antagonistic pleiotropy, in which the benefits of early reproduction outweigh the costs of early aging.

Some of the responses to early developmental conditions are harder to interpret. Most nephrons are formed in the last trimester of pregnancy. Babies who are born premature or who have low birth weight because of intrauterine growth retardation have a reduced number of nephrons at birth and so are at increased risk of developing chronic kidney disease and hypertension as they age and lose renal function (Hershkovitz et al. 2007). Reduced nephron number may be part of an evolved life history strategy to conserve resources for reproduction rather than invest in physiological capital. Alternately, it may simply be the bad outcome of developing in an unhealthy intrauterine environment. In either event, the relationship between nephron number at birth and the risk of renal disease in adult life highlights the importance of physiological capital in prolonging health.

While our developmental plasticity must have evolved because on balance it increased the reproductive fitness of our ancestors, this plasticity renders us susceptible to environmental toxins that alter our life history strategies. There is currently much concern about the possibility that endocrine disruptors, exogenous compounds that block or mimic the actions of hormones, are causing early puberty and are increasing the risk of disease in adult life (Schug et al. 2011). Since these agents were not part of our ancestral environment, our responses to them are pathologies, not adaptations.

The epidemiologist David Barker and others have collected a wealth of epidemiologic evidence documenting that babies who are undernourished during fetal development or early infancy are at increased risk of developing diabetes, cardiovascular disease, and other diseases in adult life (Barker 2004). Barker's work has led to the creation of a new field, developmental origins of health and disease, which is devoted to exploring the relationships between early development and adult disease (Gluckman et al. 2007; Gluckman et al. 2010). These relationships are complex and hard to study, especially in humans, and our life histories are so different from the life histories of model organisms that research on these organisms may not be directly applicable to us. The most common measures of fetal development, gestational age and birth weight, are probably not good proxies for the nutritional or endocrine cues that actually affect our life histories. We are just beginning to identify the relevant signals and the developmental periods during which they act, and the epigenetic mechanisms by which they modify gene expression, hormone secretion, and development (Drake et al. 2012).

Barker and Nicholas Hales have suggested that the increased risk of diabetes among people who were undernourished early in development is a tradeoff for enhanced fetal and neonatal survival (Hales and Barker 2001). Specifically, they suggested that a poor nutritional environment during development leads to insulin resistance, which slows growth and preserves nutrition for brain development and reproduction but which subsequently increases the risk of developing diabetes. If their hypothesis is correct, the physiological response to early environmental deprivation would be another example of epigenetic antagonistic pleiotropy.

The risk of developing diabetes and cardiovascular disease is particularly great for people who are small at birth and infancy but who subsequently gain excess weight in childhood. Increased adiposity exacerbates insulin resistance and may also increase inflammation in these "thin then fat" people (Hales and Barker 2001). While the pathophysiological mechanisms are still being worked out and our current understanding is far from complete, these observations have called attention to the problems that can result from a mismatch between the environment a fetus or infant experiences during development and the environment it then encounters later in life (Gluckman and Hanson 2006). Changes in food supplies in economically developing countries and among people who migrate from developing to developed countries may be exacerbating this mismatch and may be contributing to the increasing prevalence of diabetes and cardiovascular disease in these populations (Wells 2010).

6

Cancer

6.1 Introduction

Cancer is understandably one of the most feared diseases in our society. Not only is it a major cause of mortality, accounting for over 500 000 deaths/year in the United States and more than 7 million deaths/year worldwide but, with a few notable exceptions, it has proven to be singularly refractory to therapy. Cancer is a family of diseases that are characterized by the abnormal or unrestrained replication of somatic cells. This unrestrained cell replication often leads to palpable swellings or masses, which may have been called cancers (<L. *cancer*, crab) because they are sometimes surrounded by swollen veins that were thought to resemble a crab's claws (Mukherjee 2010). Cancers may arise from many different cell types in many different tissues. Most of the lethal cancers in humans are carcinomas (<Gr. *karkinos*, crab, plus -*oma*, tumor), which arise from epithelial tissues. Cancers of the lung, colon, breast, and prostate are the major lethal cancers in the United States. Cancers that arise from mesenchymal tissues (bone, cartilage, muscle, connective tissue) are known as sarcomas, while those that arise from immune or blood-forming tissues are leukemias or lymphomas. Because a diagnosis of cancer has been thought to carry a social stigma as well as a death sentence, cancers have been called by a number of synonyms and euphemisms—tumors, growths, neoplasms, or simply "ca" (Sontag 1978). Tumors are frequently classified as benign or malignant; benign tumors are usually surrounded by a connective tissue capsule and remain localized to their site of origin, whereas malignant tumors invade surrounding tissues and/or migrate (metastasize) to distant parts of the body. Cancers are malignant tumors.

6.2 Cancer as a disease of aging

Cancer is largely a disease of the elderly. The incidence of most cancers rises steeply with age and more than half of all cancers are detected in people over 60 (Vogelstein and Kinzler 1993). (A few types of cancer, and especially leukemias and lymphomas, are important causes of childhood mortality. Childhood cancers differ from adult cancers in that they are frequently associated with inherited germline mutations that predispose children to these diseases.) On the other hand, the incidence of and mortality from most cancers peak at around age 80 to 85 and decline at older ages (Harding et al. 2012). This decline in cancer mortality is puzzling. It is presumably related to the stabilization or decrease in

Evolution and Medicine. First Edition. Robert L. Perlman © Robert L. Perlman 2013.
Published 2013 by Oxford University Press.

age-specific mortality rates in older people that we discussed in Chapter 2. Cancer is a disease of aging from a life history perspective as well as in terms of its epidemiology. Cancers result from unrepaired cellular damage that leads to the breakdown of the cellular and physiological mechanisms that normally regulate and restrain cell growth. We must understand these normal regulatory mechanisms before we consider how their malfunction may lead to cancer.

6.3 Regulation of cell growth and replication

The growth of differentiated cells in multicellular organisms like ourselves is tightly controlled. The 10^{13} (10 trillion) or so cells in our bodies are derived from a single fertilized ovum. Except for the mutations that may accumulate during the replication of somatic cells, these cells are genetically identical. They differ phenotypically because of the epigenetic modifications we discussed earlier, including DNA methylation, chemical modifications of histones and other chromosomal proteins, and the actions of regulatory RNA molecules. These epigenetic modifications are typically introduced early in development and are generally heritable during the mitotic replication of somatic cells. Different epigenetic modifications in different stem cell populations produce cell lineages with distinct patterns of gene expression and distinct specialized phenotypes (neurons, muscle cells, etc.). Multicellular organisms may be thought of as associations of genetically identical but epigenetically distinct cell populations that cooperate for survival and reproduction (Buss 1987). In metazoans, or multicellular animals, only the germ cells give rise to new organisms and pass on their genes to these offspring. Bodily or somatic cells transmit their genes indirectly, through the germ cells to which they are genetically related. Somatic cells enhance their inclusive fitness by promoting the reproductive fitness of the organisms of which they are a part. The evolution of metazoans has entailed selection for mechanisms that regulate the replication of somatic cells and resist or counter the unrestrained growth of these cells. Cell growth and replication are regulated both by intrinsic intracellular mechanisms and by ecological interactions between cells and the extracellular fluids, extracellular matrix proteins, and neighboring cells that constitute their environment.

Much of our knowledge about the biology of cancer cells comes from the study of cell growth in vitro, in cell culture. The cell cycle, one round of cell division from the time a new cell is born until the time it divides into two daughter cells, is conventionally divided into several distinct phases: the synthesis of cellular components and an increase in cell size (the Gap1 or G1 phase); DNA replication (the S phase); another "Gap" period after DNA replication, in which errors in replication are repaired (G2); and, finally, mitosis, the segregation of chromosomes into two separate nuclei followed by cytokinesis, the division of the parent cell into two daughter cells (M). Cells that stop replicating enter a quiescent phase known as G0, where they may stay temporarily or permanently (Figure 6.1).

Because proper functioning (and quiescence) of the cell cycle is critical for cell survival and replication, the cycle is closely and carefully controlled. Regulation of the cell cycle is dauntingly complex. In simplified form, we may think of the progression of cells through the

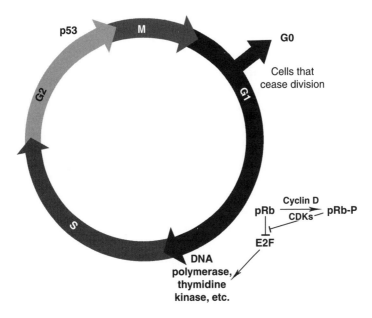

Figure 6.1 The cell cycle.

cell cycle as regulated by a series of molecular switches. These switches are made up of pairs of proteins known as cyclins and cyclin-dependent kinases, or CDKs. (Cyclins are so named because their concentrations vary during the cell cycle.) When bound to the appropriate cyclins, CDKs catalyze the phosphorylation of specific target proteins, which in turn regulate the passage of cells through the cell cycle. Control of the onset of DNA replication (progression from G1 to the S phase of the cell cycle) exemplifies this general process. In nondividing cells and during G1, a protein known as the retinoblastoma protein, or pRb, binds to and inhibits the activity of the E2F family of transcription factors, which are required for the synthesis of enzymes such as DNA polymerase and thymidine kinase, which in turn are necessary for DNA replication. The gradual accumulation of cyclin D during G1 leads to the activation of specific CDKs (CDK4 and CDK6), which phosphorylate and inactivate pRb. Phosphorylation of pRb relieves the inhibition of E2F, increases the synthesis of enzymes required for DNA replication, and ultimately allows DNA synthesis to begin (Suryadinata et al. 2010). Regulation of the transition from G1 to S is more complex and more tightly controlled than this simple linear pathway may imply. CDK4 and CDK6 phosphorylate other proteins in addition to pRb, and pRb probably has other activities in addition to inhibiting E2F. Nonetheless, this simple scheme highlights some of the key steps in the transition from G1 to S and the initiation of DNA replication. Phosphorylation of other proteins by other cyclin-CDK complexes regulates the initiation of other steps in the cell cycle.

Cell division requires that all steps in the cell cycle be carried out properly and in the correct sequence. Cells have a number of mechanisms to recognize and respond to damage or mistakes in the cell cycle. One important set of mechanisms involves the recognition of mistakes that inevitably occur during DNA replication and the correction of these mistakes before cells enter the M phase of the cell cycle and divide, thereby minimizing the production of somatic mutations. Even in the absence of exogenous mutagens, bases in DNA can become oxidized (recall the oxidation of cytosine residues we discussed earlier) or undergo a variety of other chemical modifications that can increase the likelihood of mistakes when the DNA is replicated. While most of this damage is repaired, correction is not complete and there is a low steady state level of abnormal bases in our genomes. In addition, a variety of mistakes can be made during DNA replication, from simple mismatches to insertions, deletions, duplications, strand breakage, and chromosomal translocations. As we discussed earlier, we devote a significant amount of energy and resources to prevention and detection of DNA damage, the processes of DNA repair, and protection against the untoward consequences of unrepaired or damaged DNA. In addition to the many enzyme systems that detect and correct mistakes in DNA replication, our somatic cells have a protein called p53 that accumulates in cells that have DNA damage (unpaired or broken chromosomes, etc.). p53 prevents these cells from continuing through the cell cycle until this damage is repaired or causes the damaged cells to undergo programmed cell death, or apoptosis.

The growth and replication of single-celled organisms, such as bacteria and yeast, is controlled primarily by the availability of nutrients. These cells replicate as rapidly as their nutrient supply permits. When they exhaust their food supply, they become dormant or die. As the French biologist François Jacob has written, "A bacterium continually strives to produce two bacteria. This seems to be its one project, its sole ambition" (Jacob 1976, p. 271). If Malthus (1798) had known about bacteria, this behavior is just what he would have predicted. In contrast, the replication and differentiation of mammalian cells is controlled not only by the availability of nutrients but also by hormones and growth-regulatory factors that modulate cell growth in response to the physiological needs of the organism. As is well known, the uptake of glucose into many somatic cells is dependent upon insulin or insulin-like growth factors. Glucose uptake is essential for the synthesis of many of the cellular constituents required for cell division. Most normal cells do not replicate in cell culture and presumably do not replicate in vivo unless they are stimulated by the appropriate growth factors. Some growth factors promote cell growth by affecting the cyclins or CDKs that regulate progression through the cell cycle, while others increase cell growth by different mechanisms. Cell growth and replication are regulated by growth inhibitory signals as well as by growth promoting factors. Some of these growth inhibitory signals, such as transforming growth factor β1 (TGFβ1), which we discussed earlier in connection with cystic fibrosis (Chapter 4), are soluble proteins, while others are components of the extracellular matrix. Different cell types are responsive to different growth promoting and growth inhibitory factors. A mix of these molecular signals modulates the replication of specific cell types, again in service to the physiological needs of the organism. Many types of cells lose the capacity to divide when they become fully differentiated. Most tissues, however, contain populations of stem cells that can replicate to replace nonfunctional or dying cells. The cells on the surface of the skin, for example, are continuously being sloughed off into the air and are replaced by new cells coming from the replication of so-called basal cells that reside just beneath the skin surface.

Extracellular matrix proteins or other components of the extracellular environment not only regulate cell division but also restrain or control cell migration. Except for the handful of specialized cell types that circulate in blood or migrate in tissues, most somatic cells remain localized in or near the area in which they were born.

Even when they are stimulated by growth factors, most normal cells have a finite replication potential in culture. One reason for this limitation arises from the process of DNA replication itself. Telomeres, the ends of chromosomes in eukaryotic cells, are made up of tandem repeats of short nucleotide sequences. Telomeres in human cells comprise hundreds of copies of the hexanucleotide TTAGGG at their 3′ ends. Because the enzymes that catalyze DNA replication are unable to replicate the 3′ ends of DNA, the telomeres become shorter—that is, they have fewer repeating TTAGGG sequences—with each round of cell division. Recall that shortening of telomeres is one of the somatic changes associated with aging. Eventually, when their telomeres get too short, chromosomes become unstable and may become joined to other chromosomes. p53 and other proteins recognize these aberrant fused chromosomes and cause the cells that contain them to undergo apoptosis. This process of telomere shortening gives human somatic cells a finite replication potential, perhaps on the order of 50 cell divisions (Hayflick 1980). Germ cells and embryonic stem cells express the enzyme telomerase, which elongates shortened telomeres, maintains telomere lengths roughly constant, and so renders these cells immortal (or, at least, increases the number of divisions a cell lineage can undergo before it dies). The telomerase gene becomes inactivated during the differentiation of most somatic cells. The limited replication potential of somatic cells is due at least in part to the loss of telomerase activity. The dependence of somatic cell replication on growth regulatory factors, the mechanisms that cause damaged or abnormal cells to undergo apoptosis, and the loss of telomerase activity all exemplify the ways in which somatic cells sacrifice their own reproductive potential for the benefit of the organisms of which they are a part.

6.4 Selection for cells that escape normal growth controls

Natural selection operates at many different levels of biological organization. The cells in multicellular organisms, like the organisms themselves, compete for survival and reproductive success in the environments in which they live—in this case, in the extracellular milieu of the organism. Despite the existence of multiple mechanisms that regulate cell survival and replication, there will always be selection for cells that escape these growth controls and that replicate at the expense of the organism—in other words, for cancer cells. In the 1970s, John Cairns and Peter Nowell suggested that cancers develop by a classical Darwinian process of variation and selection (Cairns 1975; Nowell 1976). Since that time, this hypothesis has been amply confirmed by a number of investigators (Greaves 2002; Merlo et al. 2006; Nunney 1999; Vineis and Berwick 2006).

Genetic mutations and epigenetic changes provide the heritable variation that is necessary for the natural selection of cancer cell populations. The mutation rate of human somatic cells is estimated to be on the order of 10^{-9} per nucleotide site per cell division (Lynch 2010). Given the size of our genomes, this means that about a half-dozen new mutations occur in each round of cell division. The spectrum of somatic cell mutations is similar to that in germline cells. Base substitutions, and especially C→T transitions, are by far the most common somatic mutations but insertions/deletions, duplications, and chromosomal rearrangements and translocations also occur. By the time people reach reproductive maturity and adult life, their replicating somatic cells may contain thousands of mutations. And by the time we reach age 60 or so, nearly every nucleotide site in our genomes is likely to have acquired a mutation in at least one cell in our intestinal epithelium alone (Lynch 2010). Epigenetic mutation rates are probably much greater than the rates of genetic mutations (Tsai and Baylin 2011). Together, genetic and epigenetic mutations provide a huge range of heritable variation among somatic cells, variation on which selection can act. As in the case of germline mutations, most of these mutations are likely to be neutral, and cells with deleterious mutations (i.e. deleterious with respect to their own survival and replication) will be eliminated by selection. But some mutant cells will have increased replicative fitness and these cells will, on average, survive and give rise to clones or lineages of daughter cells. Cancers, like other diseases of aging, result from the limitations or imperfections of somatic maintenance—in this case, from the failure to prevent or repair mistakes in DNA replication, to prevent or reverse epimutations, to kill mutant and damaged cells, etc.

A number of genes have been identified whose expression, overexpression, or mutation leads to an increased incidence of cancer. These genes, which were discovered in tumor viruses and subsequently found to be normal constituents of the human genome, are known as oncogenes (Varmus 1990). Many oncogenes encode the synthesis of growth factors, growth factor receptors, or other proteins in the signaling pathways by which growth factors stimulate cell growth. The recognition that specific genetic mutations are associated with cancer is fundamental to our understanding of carcinogenesis. Needless to say, oncogenes did not evolve because of their ability to cause cancer but rather because of their roles in normal developmental and physiological processes.

Other genes, known as tumor suppressor genes, regulate cell replication, promote apoptosis, or enhance DNA repair. The inactivation of these genes increases the probability of cancer formation or progression. The pRb gene was originally identified as a tumor suppressor gene. It got its name because it is mutated or deleted in many patients with retinoblastomas. Understandably, cells in which pRb is inactivated will no longer have the restraint on DNA synthesis that this protein normally provides. The gene that encodes p53 is another tumor suppressor gene. Cells with mutated and inactivated p53 show decreased apoptosis and thus may continue to replicate despite having damaged DNA. Many human tumors have mutations that inactivate p53. The genes that function in mismatch repair of DNA are also tumor suppressor genes. Tumor suppressor genes probably evolved because they participate in the regulation of the cell cycle and normal development, and they may protect cells against viral pathogens, not because they suppress the formation of cancers. From the perspective of evolutionary life history theory and the evolutionary biology of aging, tumor suppressor genes can be understood as genes that promote bodily repair and longevity by reducing the probability of DNA damage or preventing the replication of damaged cells.

6.5 Cancer progression

The development of cancer is a multistep process that appears to require several rounds of mutation or epimutation and selection of mutant cell lineages. Douglas Hanahan and Robert Weinberg identified what they called six "hallmarks of cancer," altered cellular phenotypes that are necessary for or contribute to the development of most types of cancer (Hanahan and Weinberg 2000; Hanahan and Weinberg 2011). These properties include independence from growth factors, insensitivity to growth-inhibitory factors, limitless potential for replication, and the abilities to evade apoptosis, to gain access to nutrients by promoting angiogenesis, and to invade tissues and metastasize. In accordance with their analysis, it appears that only a small number of mutations are required to convert a normal cell into a cancer cell. The formation of a colon cancer, for example, appears to require at least five distinct mutations but possibly not many more (Fearon and Vogelstein 1990). And the generation of some cancers, again using colon cancer as an example, goes through a recognizable progression from normal epithelium to polyp to adenoma to carcinoma, reflecting several rounds of mutation and selection. At each step in this process, clones or lineages of mutant cells will compete for survival and replication, and so will continue to undergo selection. Cells will be selected for such traits as increased replication, decreased apoptosis, avoidance of host defenses, ability to spread locally and to metastasize to distant sites, and resistance to therapy—that is, for the hallmarks of cancer.

Because cancers arise from a process of random mutation and selection, every cancer is genetically unique. Although the mutations that lead to cancer must disrupt common regulatory pathways, selection is based on cellular phenotypes, not genotypes. Different mutations may lead to the development of the hallmarks of cancer in different tumors. In a few instances, such as the chromosomal translocation associated with chronic myelogenous leukemia, a specific genomic aberration is necessary for the development of a specific type of cancer (Rowley 1973). By and large, however, no specific mutations are either necessary or sufficient for the development of any given type of cancer.

Recent advances in DNA sequencing have permitted the sequencing of multiple biopsy samples from primary tumors and their metastases in individual patients and have opened the possibility of sequencing individual cancer cells (Gerlinger et al. 2012; Navin and Hicks 2011). Although these studies are still in their infancy, they are providing new information about tumor evolution. Notably, they are revealing a high level of intratumor genetic heterogeneity. Not only may metastases differ genetically from primary tumors, but different regions of primary tumors may have different sets of mutations. Phylogenetic reconstruction of the pathway of tumor evolution shows a branching pattern with convergent phenotypic evolution, such that different mutations in different portions of a tumor may give rise to cells with similar phenotypes. Intratumor heterogeneity supports the idea that cancers are made up of multiple cell lineages that are competing for space and nutrients within the confines of the tumor.

Although only a small number of mutations are required for the development of a cancer, most cancer cells contain hundreds or thousands of mutations. Some of the mutations commonly found in cancer cells are mutations in genes involved in DNA repair or in other tumor suppressor genes, which decrease the ability of these cells to repair any subsequent damage they may undergo. Cancer cell lineages that have decreased capacity to repair damaged DNA may have "mutator phenotypes" (Loeb 2011). In addition, mutations in genes that affect DNA methylation or histone modifications may lead to many epigenetic changes in the mutant cells. Only a few of the many mutations or epimutations in cancer cells are thought to be "driver" mutations that affect critical regulatory pathways and contribute to the growth and replication of these cells. Most are assumed to be "passenger" mutations that do not play a significant role in cancer progression. These passenger mutations are carried along by a process analogous to the founder effect we discussed in Chapter 3. They are present in cancer cell lineages because they just happened to be present in the cells that founded these lineages. It is difficult to determine which mutations actually enhance the fitness of cancer cells and lead to the hallmarks of cancer phenotypes that are under selection.

Many cancers contain subpopulations of cells with the properties of stem cells. These may be normal tissue stem cells from which the tumors originated or they may be tumor cells that have dedifferentiated and have acquired the characteristics of stem cells. The significance of cancer stem cells is still controversial and may differ in different types of tumors but these cells are likely to play an important role in cancer biology. Although stem cells have a relatively slow rate of replication, which reduces the rate at which they accumulate mutations, cancer stem cells may already have accumulated a host of mutations before they assume the stem cell phenotype. (Some mutations, including C→T transitions, can also occur in the DNA of nondividing cells, so even nonreplicating or slowly replicating stem cells may acquire a significant load of mutations (Lynch 2010).) The low replication rate of cancer stem cells renders them relatively resistant to chemotherapy and radiation therapy (Frank et al. 2010). When stem cells do divide, they divide asymmetrically. One of the daughter cells replaces its mother while the other generates a clone of rapidly dividing cells. These properties of stem cells make them well suited to underlie the persistence and recurrence of cancers and the production of new precancerous and cancerous cell lineages. Cancer stem cells may be the source of the proliferating cancer cells that replace the dying cells in tumors.

Clonal competition among cancer cell lineages will lead to selection for lineages that devote most of their energy to cell growth and replication at the expense of cellular maintenance. The decreased DNA repair that characterizes many cancers is an example of the reduced cellular maintenance in successful cancer cell lineages. On the other hand, the production of cancer stem cells represents utilization of resources for maintenance of the lineage. Cancer cells may enhance their inclusive fitness by promoting the formation and survival of the stem cells in their lineage. Evidently, cancer cell lineages evolve their own life history strategies.

6.6 Ecology of cancers

The focus of cancer research on the genetic evolution of cancer cell populations has tended to draw attention away from the fact that this evolution takes place in the environments in which the cells are replicating (Merlo et al. 2006). There is now increasing recognition that cancers are complex structures comprising many different cell types. In addition to multiple cancer cell lineages, cancers contain endothelial and other vascular cells, fibroblasts, and cells of the immune system. Some of the noncancerous cells in cancers may be part of our defenses against the growth and spread of cancers but others may be recruited by the cancer cells and contribute to cancer progression. Moreover, cancers are embedded in the normal tissues in which they are growing. The ecological interactions among the different cell types in the cancers and between the cancers and the surrounding normal tissue play a critical role in the natural history of cancers. These environments determine the fitness of competing cancer clones. Although we may talk about cancer cells replicating autonomously, they are of course still dependent upon their environment for the nutrients and other factors they need to grow and replicate and they may still be influenced by environmental factors that limit their survival or restrain their replication. Recognition of the importance of the environment in the progression of cancers forms the basis of some approaches to cancer chemotherapy, such as the use of angiogenesis inhibitors.

The cells that make up cancers are functionally as well as genetically heterogeneous. In addition to competing, the cells or cell lineages in cancers may cooperate in ways that support the growth and maintenance of the tumor. For example, some cells in tumors may excrete lactate while other cells take up and metabolize lactate. Some cancer cells may promote angiogenesis while others may help cancers escape immune defenses. Because of this functional cooperation between phenotypically different cells, cancers are sometimes referred to as organs or even as superorganisms (Grunewald et al. 2011).

Just as organisms modify their environments, so cancer cells modify theirs. The metabolism of cancer cells may cause the environment around the cells to become hypoxic and acidic, which in turn will lead to selection for cells that can grow and replicate in this altered environment. In addition, cancer cells may secrete substances that alter the properties of the noncancerous cells in tumors. Cancers are evolving ecosystems.

Even the process of metastasis is constrained by the interactions of cancer cells with their environment. Metastases do not occur randomly. In the nineteenth century, Stephen Paget developed the "seed and soil" hypothesis to account for the observation that certain cancers are able to grow only in the environments provided by certain tissues (Paget 1889). We are just beginning to learn the ecological features that make some tissues receptive to the growth of metastatic cancers.

6.7 Anti-cancer defenses

Cancers occur in many metazoan species. Although cancers may be especially prevalent in vertebrates because of our relatively large sizes and long lives, they have also been reported in arthropods and mollusks (Leroi et al. 2003). Plants develop localized tumors but apparently do not have malignant or metastatic cancers. Perhaps the rigid plant cell walls and extracellular matrix, together with the rudimentary circulatory systems in plants, prevents the spread of neoplastic cells.

The mechanisms discussed earlier, which repair mistakes in DNA replication, cause damaged cells to undergo apoptosis, and reduce mutation rates, now function as defenses against cancer but probably evolved early in the history of eukaryotes, before cancers arose. Vertebrates have evolved a variety of mechanisms that decrease the fitness of precancerous and cancerous cells—in other words, decrease the selective advantage of these cells—and so counter the growth and spread of tumors. These mechanisms are likely to have evolved as anti-cancer defenses. Many incipient cancers become surrounded by connective tissue capsules. These capsules decrease the supply of nutrients to the tumors and hinder them from invading host tissues. Both the extracellular matrix and normal cells in the environment of cancer cells appear to restrain the growth of many incipient cancers (Bissell and Hines 2011). In addition, immune recognition of cancer cells may lead to apoptosis of these cells.

Many tumors appear to be dormant and do not grow for long periods of time. Progression of large colonic adenomas to colon cancers may take as long as 17 years (Jones et al. 2008). But the apparent dormancy of tumors is deceiving, because the cells in these nongrowing tumors are continuing to divide and to accumulate mutations. The tumors don't grow because their rate of cell division is balanced by their rate of cell death. The long dormancy of many tumors is consistent with the hypothesis that multiple steps, again either genetic mutations or epimutations, are required to create clinically significant cancers. An initial mutation leads to the production of a clone that remains small and localized until another mutation enables a new lineage to grow faster and spread. Cancers, or incipient cancers, appear to be constantly arising but our anti-cancer defense mechanisms prevent most of these cancers from growing and spreading. By they time they get to be 80 years old, for example, most men have what appear to be cancers in their prostate glands. The overwhelming majority of these tumors remain microscopic and don't cause any clinical problems; they are discovered only by prostate biopsy or at autopsy. But while natural selection has provided us with powerful anti-cancer defenses, it has not made these defenses perfect. As evolutionary life history theory would predict, we have invested enough energy in anti-cancer defenses to keep most of us from succumbing to cancer before we have finished our reproductive and child raising lives. Although cancer is a major human disease, it represents the relatively infrequent failure of our evolved defense mechanisms to prevent the formation of mutant cells or to reduce the fitness of these mutant cell populations and so contain their growth.

6.8 Carcinogenesis and cancer prevention

As we have seen, cancers result from unrepaired somatic mutations and epimutations. There are families who carry mutations in pRb or other tumor suppressor genes and who are at increased risk of developing cancer but germline mutations play only a small role in the epidemiology of cancer. Many of the clinically important cancers, including those of the lung, colon, and breast, arise in epithelial tissues that undergo cell replication and clonal expansion throughout life. This continuous cell replication increases the accumulation of the somatic mutations that are necessary for cancer. Leukemias and lymphomas arise in cells which have specialized mechanisms that increase rates of mutation and genetic recombination. Mutations and genetic recombination are required for the development of the immune system. Our susceptibility to leukemias may be understood as a tradeoff for our adaptive immunity.

Environmental factors strongly affect the rate of somatic mutation and so are important risk factors for cancer. The incidence of somatic mutations, and therefore of cancer, will be determined by the rate of cell division in a tissue and by the balance between rates of DNA damage and DNA repair in this tissue (Ames et al. 1993). Thus, in general, there are two types of carcinogens—growth-promoting factors, which increase the rate of cell division, and mutagens, which increase the extent of DNA damage per cycle of cell division. Although not all carcinogens are mutagens, all mutagens appear to be carcinogens. The high incidence of breast cancer in the United States and other developed countries illustrates the importance of cell replication in the development of cancer. During each menstrual cycle, estrogens stimulate the replication of breast cells. Throughout most of human history, women are likely to have had relatively few menstrual cycles; for the majority of time between menarche and menopause, women were either pregnant or nursing, or they were too poorly nourished to ovulate. Given our good health and nutrition, and our low fertility rates, women in developed countries now have many more menstrual cycles than did our ancestors and so they are exposed to many more rounds of estrogen stimulation. Simply by increasing cell replication, estrogens increase the probability that breast cells will accumulate mutations that ultimately lead to breast cancer (Eaton et al. 1994).

Chronic inflammation may also increase cell replication and lead to cancer. Colonization of the stomach with the bacterium *Helicobacter pylori* is a risk factor for the development of gastric cancer. Bacterial infection evokes an inflammatory response that increases the replication of gastric epithelial cells, which in turn increases the risk of cancer (Moss and Blaser 2005).

Agents that increase rates of cell division, such as estrogens and *H. pylori*, can increase the incidence of cancer without directly affecting the processes of DNA damage and repair. Other environmental agents interact with DNA and so increase the frequency of errors in DNA replication. (The effect of *H. pylori* is complicated because inflammation not only increases the replication of gastric epithelial cells but may also increase oxidative damage to the DNA in these cells.) Because repair of these errors is incomplete, these agents are mutagens—they increase the rate of formation of mutations. Ultraviolet light is the most ubiquitous environmental mutagen. UV-induced mutations are responsible for the high incidence of skin cancers. Other environmental mutagens include X-rays and other types of radiation, tobacco, etc. Reduction in smoking has undoubtedly been the most important public health advance in the last 50 years (Peto et al. 2000; Stewart et al. 2009). Although lung cancer is still an important cause of death, it is much less important than it was before the anti-smoking campaigns of the 1950s and 1960s. Reduction of exposure to environmental mutagens remains a most valuable strategy for reducing the incidence of cancer.

Oncogenes were discovered in viruses that cause tumors in birds and in mice. Viruses seem to be especially important causes of cancer in these other species but are thought to be responsible for only about 15–20% of human cancers. The best known oncogenic virus that infects humans is human papilloma virus (HPV), which is a major risk factor for both cervical and oral cancers (zur Hausen 2002). HPV is a DNA virus whose genome becomes integrated into the genomes of cervical epithelial cells and the other cells it infects. The virus has evolved mechanisms that stimulate the replication of cervical epithelial cells. One viral protein binds to and inactivates the pRb protein and so speeds up the initiation of DNA synthesis, while another protein interacts with p53 and prevents p53 from causing apoptosis or blocking the onset of mitosis. The increased replication of cervical cells enhances viral replication and so promotes transmission of the virus. But the increased replication of cervical cells is accompanied by the formation of mutant and damaged cells, which may go on to become malignant. Cervical cancer does not benefit HPV. Cancer does not increase the replication or transmission of the virus. It is a tragic but incidental byproduct of selection for fast-replicating, HPV-infected cervical cells. Most cases of primary liver cancer, hepatocellular carcinoma, appear to be due to hepatitis viruses, specifically the hepatitis B and C viruses. Most if not all oncogenic viruses in humans are sexually transmitted viruses. As we shall discuss later, sexually transmitted pathogens are under strong selection to evade host defenses and cause persistent infections (Ewald and Swain Ewald 2012).

To date, most of the attention given to cancer prevention has focused on reducing mutations; less attention has been given to factors affecting selection for mutant precancerous cell lineages. The multistep nature of cancer progression blurs the distinction between cancer prevention and cancer therapy. As we have discussed, there is increasing evidence that the fitness of cancer cells and the growth of cancers are constrained by the ecological interactions between these cells and the tissues that surround them (Bissell and Hines 2011), and there are intriguing suggestions that interventions may alter these interactions in ways that affect selection for cancers (Vineis and Berwick 2006). Moreover, clonal competition among cancer cell lineages may constrain the growth of cancers. Clonal competition may have important implications for the clinical management of cancers. For example, if one lineage is sensitive to chemotherapy while another lineage is resistant, treatments that kill cells in the sensitive lineage may simply remove the competition and open up space and resources that enable the resistant lineage to expand. Cytostatic drugs, drugs that slow or stop cell division, may be more effective therapeutic agents than cytotoxic drugs because they don't remove the competing lineage and, in contrast to many cytotoxic drugs, they are not mutagenic (Greaves and Maley 2012). Better understanding of the ecology of incipient cancers may lead to strategies that reduce selection for the growth and spread of these cancers.

Given our evolutionary understanding of cancer, it seems unlikely that this disease will ever be eliminated from the human population. Cell replication is inevitably accompanied by damage to DNA, and although we have evolved powerful mechanisms to correct this damage and to kill damaged cells, these mechanisms are not perfect. Informed by this evolutionary understanding, our cancer prevention efforts should continue to be focused on reducing our exposure to mutagens or to agents that unnecessarily stimulate cell division, and on developing ways of enhancing DNA repair and decreasing selection of mutant precancerous and cancerous cells.

7

Host–pathogen coevolution

7.1 Introduction

Our bodies provide environments in which a multitude of other organisms can live and reproduce. Our gastrointestinal, respiratory, and reproductive tracts, and our skin, are normally colonized by communities of microorganisms, which are known collectively as the human microbiome. The microbiome is made up of thousands of bacterial species, which together comprise roughly ten times as many cells (approximately 100 trillion vs 10 trillion) and contain more than 100 times as many genes (roughly 3 million vs 21 thousand) than are present in "our own" bodies. Our coevolution with the organisms in our microbiome has resulted in fitness benefits to both parties. We provide a habitat and nutrition for these organisms, and they enhance our fitness in many ways. Microbial metabolism of food that our own enzymes cannot digest contributes about 10% of the energy we get from our diets. In addition, our microbiome helps protect us from becoming colonized by other, potentially harmful, microorganisms. Finally, our microbiome shapes the development of our immune systems and our gastrointestinal tract. The human microbiome is the subject of active investigation (Cho and Blaser 2012; Walter and Ley 2011). There is growing understanding of the role of our microbiome in normal physiology and increasing recognition that alterations in our microbiome increase our risk of developing a variety of immunological, gastrointestinal, and other diseases.

Despite the many benefits we get from our microbiome, some of the organisms in the microbiome do on occasion cause disease. More importantly, many exogenous disease-causing organisms, or parasites, can infect us. Ecologists often distinguish between microparasites (viruses, bacteria, and fungi), which not only grow in us but also reproduce inside us, and macroparasites (helminths, arthropods, and protozoa), which grow in us but do not reproduce and multiply inside us (Anderson and May 1991). In medicine, we usually refer only to this latter group as parasites and we think about the entire group of disease-causing organisms as pathogens. Well over 1000 human pathogens have been identified (Cleaveland et al. 2001). Despite our evolved defenses against pathogens, natural selection will always favor the spread of organisms that can grow in us and be transmitted among us. Because some of these organisms will be pathogens, infectious diseases will always be part of the human condition. The theory of host-parasite or host–pathogen coevolution provides a framework for understanding the natural histories of infectious diseases.

Evolution and Medicine. First Edition. Robert L. Perlman © Robert L. Perlman 2013.
Published 2013 by Oxford University Press.

7.2 Epidemiology of pathogen transmission

Pathogens have complex life histories. The host—in our case, human—population represents a "patchy" environment for infectious agents. In addition to replicating in an individual host (a person), these organisms must be transmitted from one person to another. The populations of pathogens that are growing within a host population are known technically as metapopulations. A more familiar way of thinking about pathogen transmission is that to our pathogens, we are like islands. Transmission of pathogens from infected to uninfected people is analogous to the colonization of islands by plants and animals. A key concept for understanding the spread of a pathogen in a host population is the basic reproductive number of the pathogen, R_0 (Anderson and May 1991). R_0 is an estimate of the average number of new or secondary cases of an infectious disease that would result from the introduction of one infected person into a population of susceptible people. It may be thought of as the total number of contacts an infected person makes with susceptible people during the time he or she is infectious times the probability that a contact will lead to transmission of the pathogen. The total number of contacts is often expressed as the number of contacts per day times the duration of infectivity. In this formulation,

R_0 = (contacts between infectious and susceptible people/day) × (probability that contact leads to pathogen transmission) × (duration of infectivity, in days)

The spread of a pathogen in a host population is analogous to the population growth we discussed in Chapter 2. Consider the population of pathogens within a single infected person as an individual and the secondary infections or new people infected by this initial population as its offspring. Just as human populations won't grow unless each adult woman produces, on average, more than one daughter who survives to reproductive maturity (that is, unless TFR > replacement rate), so a pathogen will not spread in a population and produce an epidemic unless each infected person gives rise, on average, to at least one secondary infection—in other words, unless $R_0 > 1$.

R_0 is the average number of people who become infected by one infected person during the time that person is infectious. The definition of R_0 does not include the generation time of the infection, the time between the infection of one person and the appearance of secondary infections (Fraser et al. 2004). If a pathogen is spreading in a host population, the generation time will determine the rate at which it spreads.

Although R_0 is referred to as the basic reproductive number of the pathogen, it is not a property of the pathogen itself. R_0 depends on the number of contacts an infected person has with susceptible people, and so it depends on the size and density of the host population. For this reason, many infectious diseases are thought of as "crowd diseases." The pathogens that cause these diseases can spread and be maintained in dense urban populations, where contacts between people are frequent and $R_0 > 1$, but not in sparse foraging groups, where contacts are fewer and $R_0 < 1$ (Diamond 1997). As we discussed earlier, these diseases became important human diseases after the agricultural revolution and the development of cities.

One simple but useful model of the spread of pathogens in a human population considers the population as made up of three groups—susceptible, infected, and resistant individuals. As a pathogen spreads in the population, susceptible individuals may have contact with or be exposed to infected individuals, and some of these exposed individuals will themselves become infected. (Many epidemiological models consider exposed individuals as a separate group but this added complexity isn't necessary for our discussion.) With many acute infectious diseases, if infected individuals recover, they are then resistant to further infection; people progress through the sequence susceptible→infected→resistant. For convenience, we shall assume that infected individuals are themselves infectious and we shall ignore births, deaths, and migration. Imagine, then, the introduction of one infected person into a population of susceptible people. As susceptible people become infected and then resistant, the pathogen's reproduction decreases, because fewer and fewer people remain susceptible to it. The actual or net reproductive number of the pathogen, R, is the average number of secondary infections produced by an infected individual when only a fraction of the population is susceptible:

$$R = R_0 \times \text{(fraction of the population that is susceptible to the pathogen)}$$

A pathogen can continue to spread in a population as long as its net reproductive number, R, remains greater than one. If the fraction of susceptible people becomes sufficiently low, such that R falls below one, the pathogen will no longer spread. The protection of susceptible people by the immunity or resistance of the majority of the community is known as "herd immunity." Herd immunity is an important public health concept because it is used to guide vaccination programs. To protect a population against an infectious disease such as smallpox or measles, it isn't necessary to vaccinate everyone; it is necessary only to vaccinate enough people so that the net reproductive number of the pathogen falls to less than one. At that point, spread of the pathogen will decrease and the disease will eventually disappear. This was the strategy that was used to eradicate smallpox from the planet.

Pathogens such as the smallpox virus, which spread through a host population until their net reproductive number becomes less than one and then disappear from that population, cause epidemic diseases. Other pathogens persist in the host population for long periods. These pathogens, and the diseases they cause, are said to be endemic. Pathogens may be endemic when susceptible people continue to enter the population at a significant rate, either because of new births or the reversion of infected and resistant people to a susceptible state. At the endemic steady state, the net reproductive number of the pathogen $R = 1$. On average, each infected person infects one previously susceptible person and a constant percentage of the population remains susceptible. The endemic steady state of an infectious disease is analogous to a stationary or nongrowing human population, whose size remains constant because, on average, each adult woman gives birth to one daughter who survives to reproductive maturity (i.e. TFR = replacement rate).

Although this brief outline of the dynamics of the spread of infectious diseases ignores many of the complexities of pathogen transmission in the real world, it does highlight most of the important factors in this process. Again, transmission of an infectious organism from an infected person to susceptible people is determined by the number of contacts the infected person has with other people during the time he or she is infectious, the probability that a contact is susceptible, and the probability that contact with a susceptible person leads to transmission of the pathogen. We have ignored the incubation or latent period of the infection, the period between the time a susceptible person is exposed to a pathogen and the time he or she becomes infectious. We assumed that people would come into contact with each other at random and we did not consider the structure (age, sex, geographic, genetic, nutritional, socioeconomic, etc.) of the population. Inclusion of these and other real-life issues would complicate the mathematics but would not change the principles (Anderson and May 1991).

We have also not considered the biology of the pathogen, whether it is a virus, a bacterium, or a eukaryote. Knowledge of pathogen biology is essential for understanding the physiological and immunological responses to it, and of course for planning therapy. For epidemiologic purposes, however, the biology of a pathogen is of minor importance. Finally, we have not considered the routes by which pathogens can be transmitted from one person to another. Pathogens that spread by different routes might spread at different rates and have different R_0s, and the route of transmission may play an important role in host–pathogen coevolution. There are only a handful of portals through which pathogens can enter our bodies or cause infections, and only a handful of routes by which they are commonly transmitted from infected to susceptible people. Pathogens can gain access to our bodies via our skin, our respiratory, gastrointestinal, or reproductive tracts, or the placenta. Except for the placenta, these are the sites that are already colonized by our microbiome. Pathogens may be transmitted by direct skin-to-skin contact, in air droplets, or through skin puncture (by arthropod vectors, other bites, needles, or sharp objects); they may spread by fecal-oral, sexual, or intra-uterine transmission; or they may be carried on fomites, substances other than food that can harbor and transmit pathogens. We shall consider some of these routes of transmission later, when we discuss specific pathogens.

7.3 Virulence and transmissibility

Two important parameters of host–pathogen interactions are virulence, or pathogenicity, and transmissibility, or infectivity. Virulence refers to the ability of a pathogen to cause disease in and death of its hosts—morbidity and mortality, for short. Virulent pathogens sicken or kill their hosts, while benign ones cause only mild disease. The virulence of pathogens is of course a major concern of patients and their physicians. From an evolutionary perspective, we are also concerned with the effects of pathogens on the reproductive fitness of their hosts. The effect of a pathogen on fitness corresponds roughly but not exactly to the severity of disease as experienced by a patient. Most importantly, pathogens that cause debilitating illness or death after the end of our reproductive period do not cause a loss of fitness. Pathogens that cause death after a long illness cause less of a fitness loss (and may be considered to be less virulent) than pathogens that kill quickly, since infected people might have the opportunity to reproduce or nurture their children, and to lead satisfying lives, during this period.

Disability-adjusted life-years is a commonly used measure of the burden of disease. Virulence is sometimes quantified in terms of disability-adjusted life-years lost per infection (Lopez and Mathers 2006). Disability-adjusted life-years lost weights death in infancy and childhood more heavily than death later in life and so is closer to a measure of fitness loss.

Transmissibility or infectivity refers to the ability of a pathogen to be transmitted from one individual to another. They are qualitative terms for the probability that contact between an infected and a susceptible person will lead to transmission of the pathogen.

7.4 Host–pathogen coevolution: hosts evolve in ways that minimize the fitness cost of pathogens

The selection pressure that pathogens have exerted on our evolutionary ancestors is straight-forward. Our ancestors evolved traits that minimized the fitness loss caused by the pathogens to which they were exposed. To the extent that loss of fitness corresponds to virulence, these traits decreased pathogen virulence. Although the rapid replication and large size of patho-gen populations, together with their relatively high mutation rates, may have limited the abil-ity of our ancestors to evolve resistance to their pathogens, the direction of our ancestors' evolution is clear.

When a new pathogen enters a human population, it will generally become adapted to grow in and be transmitted among people with the most abundant genotypes in the population—that is, to people with the genotypes and phenotypes it encounters most fre-quently. Pathogens that are adapted to people with one genotype may not grow in or be trans-mitted as well among people with other genotypes. Therefore, people with rare genotypes may have increased resistance to the pathogens. In this way, pathogens may cause frequency dependent selection of rare genotypes and may increase the genetic diversity of their host pop-ulation (Haldane 1949a). Frequency dependent selection by pathogens is thought to be an important reason for the maintenance of polymorphisms at the major histocompatibility com-plex (MHC) loci and in other genes that play a role in resistance to pathogens.

The other aspect of host resistance is decreased infectivity. We and other multicellular organisms have a number of barriers (epithelial cell layers, mucous secretions, cilia, etc.) that separate us from the outside world and that restrict the entry of pathogens. Although these barriers probably evolved for reasons other than minimizing the transmission of pathogens, they do have this benefit. As we noted earlier, our microbiome also protects us from becom-ing infected by pathogens. This protection is based in part on the principle of competitive exclusion (Hardin 1960). If two different species or strains of the same species compete to occupy an environmental niche, one of these species will outcompete the other and exclude it from that niche. Our resident microbial species have coevolved with us and are well adapted to our bodily environments, and so they are good competitors.

Behavior plays an important role in exposing us to or protecting us from pathogens. The emotion of disgust, which is elicited by olfactory and visual as well as gustatory stimuli that are associated with infectious diseases, almost certainly evolved as a defense against patho-gens (Curtis 2011). Disgust not only motivates personal behaviors that help us avoid pathogens but also underlies the adoption of cultural practices, such as quarantine and public sewage systems, that further decrease our exposure to pathogens.

7.5 Host–pathogen coevolution: pathogens evolve in ways that optimize their fitness

Because pathogens have complex life cycles, involving both replication in an individual host and transmission among hosts, the ways in which natural selection shapes the evolution of pathogens is also complex. In the past, many people believed that pathogens would inevitably evolve towards decreased virulence. Because pathogens are not, in general, transmitted from dead hosts, people assumed that there would be selection for pathogens that kept their hosts alive and healthy. Both theoretical considerations and experimental studies have shown that this idea is mistaken. Populations evolve because natural selection promotes the spread of traits that increase the survival and reproduction of organisms in the populations. The effects of a population on its environment, including its effects on other species, are incidental consequences of selection for increased reproductive fitness of the organisms in the population. Pathogens are no exception to this rule. They too evolve in ways that optimize their own fitness, even if this results in harming or killing their hosts.

As we mentioned earlier, the population of pathogens in a host has some of the characteristics of an individual organism. Just as organisms must survive and reproduce, pathogens have to replicate within an individual host (analogous to the growth and survival of a multicellular organism) and they must be transmitted from one host to another (analogous to reproduction). Pathogen populations have life histories that are analogous to the life histories of multicellular organisms like ourselves (Perlman 2009). As pathogens replicate in a host, within-host selection will favor those pathogen genotypes that enable pathogens to survive, evade host defenses, and replicate rapidly. The effects of a pathogen on its host are likely to be related in some way to the number of pathogens in the host. Even if the pathogen does nothing other than utilize host resources for its own growth and replication, it will be depleting the host of resources the host needs for its survival. Thus, within-host selection will in general lead to increased pathogen virulence. But this is not selection for increased virulence. Rather, increased virulence is an incidental byproduct of selection for pathogen replication. Again, this is ordinary natural selection, where selection for survival and reproduction of organisms leads to incidental changes in the environment—in this case, their hosts.

Our immune defenses are a major cause of the death of our pathogens, and so successful pathogens have evolved in response to these defenses. Pathogens have evolved three broad life history strategies, based on their interactions with our immune system (Perlman 2009). Some pathogens, such as rhinoviruses (the viruses that cause the common cold), are killed by our innate immune defenses. These defenses are activated immediately upon exposure to a pathogen. Pathogens that are susceptible to our innate immune responses cause infections with latent and infectious periods of only a few days each. Although we may develop immunity to the strain of pathogen with which we were infected, we clear the pathogens and recover from these infections before adaptive immunity develops. Many other pathogens, such as the viruses that cause measles, chicken pox, and mumps, are not killed by innate immune defenses but are vulnerable to our adaptive immune responses. Our adaptive immune defenses take days to become manifest and weeks or months to become fully developed. These pathogens produce acute infections with time-limited latent and infectious periods. They are typically cleared within two to four weeks from the onset of infection. Finally, a third group of pathogens has evolved mechanisms to evade or resist the adaptive immune system. These pathogens cause chronic infections that can last from many months up to the life of the host. *Mycobacterium tuberculosis* is one such pathogen. *M. tuberculosis* is often transmitted from reactivated infections decades after the initial infection.

Pathogen fitness depends upon transmission to new hosts as well as on growth in individual hosts. As pathogen populations are replicating in different hosts, among-host selection will select those pathogen genotypes that are most efficiently transmitted to uninfected hosts. Just as the traits of multicellular organisms are determined by tradeoffs between growth, somatic maintenance, and reproduction, so the traits of pathogens are determined by the tradeoffs between within-host and among-host selection, or between growth and survival within a host and transmission among hosts. The life histories of pathogens, like those of other organisms, are shaped by tradeoffs and constraints.

The basic reproductive number of a pathogen, R_0, is a good measure of fitness for pathogens that cause endemic diseases. Pathogens that spread most efficiently in the host population are the ones that will survive in that population. A reasonable expectation is that endemic pathogens will evolve to maximize R_0 (Anderson and May 1991). This selection will increase the number of contacts an infected person has with susceptible people, either by increasing the duration of infectivity or by affecting the behavior of the infected person, and the probability that a contact will lead to transmission of the pathogen. It is harder to specify a fitness measure for pathogens that cause epidemic diseases. Recall that the definition of R_0 does not include the generation time of an infection, the time between the infection of one person and the infection of the contacts to whom he or she has transmitted the pathogen. In epidemic diseases, when the number of infected hosts is increasing, there will be selection for early transmission as well as for increasing R_0. Pathogen fitness will be a function of R_0 and the generation time of the infection. Selection for early reproduction of epidemically spreading pathogens is analogous to the selection for early reproduction in expanding human populations. Despite this complication, we can consider R_0 a generally useful measure of pathogen fitness.

Because of the constraints on pathogen evolution and the tradeoffs between infectivity and virulence, the outcome of selection for pathogen fitness is different for each host-pathogen interaction. Genes that enhance infectivity may be pleiotropic and may either increase or decrease virulence. For some pathogens, transmissibility is positively correlated with virulence. As pathogens replicate and increase in number in a host, not only may the host become more debilitated but the transmission of these pathogens to new hosts may become more efficient. Selection for increased fitness of these pathogens will result in increased virulence. On the other hand, if pathogens are so virulent that they kill their hosts before the hosts have had time to contact uninfected individuals, they will undergo selection for decreased virulence. This is one explanation for why some of the exotic and frightening pathogens that are discussed in popular media, such as Ebola virus, do not cause major epidemics. These pathogens kill infected people so rapidly that their $R_0 < 1$ and so they don't spread in human populations. Such pathogens will become much more dangerous threats to public health if they evolve traits that increase their R_0. An increase in R_0 would most likely be due to the evolution of traits that reduce the rate at which these pathogens kill the people they infect, so that these people will have more opportunities to transmit their infections. In other words, before these pathogens can cause human epidemics, they will have to evolve decreased virulence.

Another way to think about the relationship between virulence and infectivity is that, in general, pathogens that are more efficiently transmitted from healthy people than from sick people will evolve to low virulence, whereas pathogens that are more efficiently transmitted from sick people will evolve to high virulence (Ewald 1994). The relative efficiency of pathogen transmission from healthy and sick hosts depends in part on the mode of transmission of the pathogen. Recall that pathogen transmission is determined by the number of contacts an infected person has with susceptible people times the probability that such a contact will lead to transmission of the pathogen. Pathogens that are transmitted directly from person to person, as in air-borne droplets, are often benign, because healthy people will have more contacts with other people and so will transmit their infections to more people than will sick or debilitated hosts. This is one explanation for why viruses such as rhinoviruses, the viruses that cause the common cold, are benign. Infected people are not very sick and so they go about their business sneezing on and transmitting their viruses to susceptible persons. Mutant viruses that had increased virulence would be selected against because people who were infected by these viruses would stay in bed sneezing on themselves. In contrast, pathogens that are transmitted indirectly, either by arthropod vectors or via fecal-oral transmission, may evolve to high virulence, because these pathogens are transmitted effectively from sick hosts. Again, these are general principles that have to be modified or adjusted for each specific host–pathogen interaction.

This, then, is the basic theory of host–pathogen coevolution. Hosts will evolve to minimize the fitness cost (or, roughly, the virulence) of pathogen infections, while pathogens will evolve to maximize their own fitness, which usually means evolving traits that maximize their basic reproductive number, R_0. Pathogen virulence is not directly selected for but is a byproduct of this coevolutionary process. In anthropomorphic terms, our pathogens don't care what they do to us; their effects on us reflect the outcome of selection acting on our pathogens and on us to increase both their fitness and ours.

7.6 Myxomatosis: a case study of host–pathogen coevolution

One of the best studied examples of host–pathogen coevolution is the coevolution of myxomatosis virus and its rabbit hosts (Fenner 1983). The European rabbit was brought to Australia by European colonists and sailors; by the late nineteenth century it had become a major agricultural pest. Myxomatosis viruses are endemic in other rabbit species but not in the European rabbit. Around the time of World War II, this virus was introduced into Australia in an attempt to control the rabbit population. The virus was initially extremely virulent, killing more than 99% of infected rabbits with a mean survival time of less than two weeks. Very soon, less virulent virus strains emerged and spread. Within several years, the majority of viral strains on the continent had a case fatality rate of 75–90% and infected rabbits had a mean survival time of 2½ to 4 weeks. Clearly, the virus had evolved to an intermediate but still high level of virulence, and it has now maintained this virulence for decades. Myxomatosis virus is a poxvirus; like other poxviruses, it causes skin lesions in infected animals. The virus is transmitted by biting insects, mainly mosquitoes and fleas. Transmission depends on the density of viruses in the skin and the duration of infectivity. Transmission of low-virulence viruses is limited by the relatively low numbers of virus particles in the skin, while transmission of highly virulent viruses is limited by the short survival time of infected rabbits. The viral strains with intermediate virulence spread and became abundant because they are associated with the highest levels of transmission—in other words, because they have the highest R_o. At the same time as the virus was becoming less virulent, the European rabbits in Australia were evolving increased resistance to the virus. The evolution of resistance was incomplete, perhaps because of the limited genetic diversity in the rabbit population. Studies of the coevolution of myxomatosis virus and rabbits provide strong evidence in support of the theory of host–pathogen coevolution outlined above. There is every reason to believe that the same principles apply to the coevolution of humans and our pathogens.

7.7 Complexities in host–pathogen interactions

The theory of host–pathogen coevolution helps to make sense of a great deal of information about the natural histories of infectious diseases. However, the generalizations presented earlier provide only a simplified overview of the complex ecological relationships between hosts and pathogens, and of the relationships between virulence and transmissibility. Many pathogens cause disease in organisms that are not their "normal" hosts, in species with which they did not coevolve. A number of pathogens that infect humans are not transmitted among humans. One example is *Clostridium tetani*, the organism that causes tetanus. *C. tetani* is a soil bacterium. It presumably produces its toxin because the toxin immobilizes or disables some other soil organisms, which then provide food for the bacteria. If a person gets a puncture wound that is contaminated by *C. tetani*, the bacteria will grow, secrete tetanus toxin, and cause a virulent infection but they will not be transmitted to other people. The fact that a toxin directed against other organisms is effective against humans reflects the common descent and biochemical similarities of humans and the intended target species. The virulence of a *C. tetani* infection in humans can't be rationalized or understood in terms of a tradeoff between virulence and infectivity because *C. tetani* did not coevolve with humans. The virulence of the bacteria in humans is just the unfortunate outcome of a trait that was selected in a different evolutionary context.

New infectious diseases arise from a combination of ecological and evolutionary changes. They may arise because humans are now being exposed to pathogens that have been adapted to other species or because of genetic changes in the pathogens that enable them to grow in and be transmitted among humans (Schrag and Wiener 1995). Of course, these factors are not mutually exclusive. As new pathogens enter the human population from other species, they will undergo selection for increased R_0 in humans. Lyme disease is due to ecological changes, to humans invading habitats where field mice and deer, and their associated ticks and bacterial pathogens, live. The virulence of these new diseases again is due to the homology between humans and the natural hosts of these pathogens. It can't be understood in terms of a trade-off between virulence and infectivity in humans because the pathogens have not (yet) evolved to be transmitted among human hosts.

Although virulence is often a byproduct of selection for pathogen fitness, pathogens may be virulent for a number of other reasons. Many pathogens, like *C. tetani*, produce toxins that harm or kill their hosts. The virulence of these organisms may result from the action of these toxins rather than from within-host competition for replication in the host. Pathogens may be virulent because they grow in and disrupt the function of critical organs. Sometimes, growth of pathogens in critical organs enhances pathogen transmission. Rabies virus grows in the central nervous system and in the salivary glands. Growth of the virus in the brain leads to paralysis, coma, and death—not just in humans, which are coincidental hosts, but also in dogs and other species with which rabies has coevolved. Viruses in the central nervous system are not themselves transmitted to other animals but these viruses increase aggressive behavior and interfere with swallowing, which enhances transmission of the virus in saliva.

In the phenomenon sometimes referred to as short-sighted or dead-end evolution, within-host selection leads to the growth of pathogens in critical organs and to an increase in virulence even though it doesn't contribute to infectivity (Levin and Bull 1994). The natural history of poliovirus infections is often cited as an example of dead-end evolution. The poliovirus infects intestinal epithelial cells and is transmitted from infected to susceptible people by fecal-oral transmission. These infections are usually benign. Occasionally, the virus escapes from the intestine and infects motor neurons; when it does, it can cause paralysis and death. But the viruses that get out of the intestine and infect the nervous system are not transmitted to other people. Infection of the nervous system is an evolutionary dead end. Moreover, the paralysis or other symptoms caused by infection of the nervous system does not increase transmission of the virus from the intestine. As far as we know, the polioviruses that infect the nervous system are genetically identical to those in the intestine. There isn't selection for neurotropic viruses. Infection of motor neurons simply represents the ability of the virus to take short-term advantage of an available ecological niche or habitat. Many of the bacteria in our microbiome are opportunistic pathogens. As long as they live on the skin or on epithelial surfaces, they are benign. If the skin or epithelia becomes damaged, the bacteria can enter our bodies and cause virulent infections. These are short-sighted infections that transiently increase bacterial replication but don't usually increase transmission of the bacteria to new hosts. Most intrauterine infections are also dead-end infections. A few pathogens are able to cross the placenta and infect fetuses but for the most part these pathogens are not transmitted from infected infants after birth.

The virulence of pathogen infections may also be caused by the immunological responses of the host or by secondary infections. Immunological responses to streptococcal infections may lead to rheumatic fever or glomerulonephritis. Influenza virus causes inflammatory changes in the lungs that predispose infected people to get bacterial infections. Most of the deaths from influenza virus result from these secondary infections. In these situations, too, pathogen virulence is not the result of selection for pathogen fitness. Again, each host–pathogen interaction is unique and results in a unique coevolutionary process. Understanding the natural history and virulence of infectious diseases requires understanding the specific features of these interactions.

Host–pathogen interactions are further complicated by the heterogeneity of the host population. Heterogeneity in age, sex, genetics, nutritional status, or other characteristics may result in variance in both virulence and infectivity. Just as pathogens adapt to be transmitted among individuals with the most abundant genotypes in the host population, they may also adapt to be transmitted among some other host subpopulation. Some of the so-called childhood diseases, such as measles and chicken pox, are typically more virulent in adults than in children. The viruses that cause these diseases have presumably become adapted to be transmitted among young children, not among adults. Their virulence in adults is probably an incidental outcome of selection for transmissibility in children and the physiological differences between children and adults.

Medical attention is understandably focused on people who are most seriously affected by pathogens. The paralysis caused by poliovirus, the encephalitis caused by measles virus, and the rheumatic fever caused by streptococcal infections are vitally important to patients, their families, and their communities. Fortunately, these life-threatening complications are relatively rare. Moreover, since they don't contribute to onward transmission of the pathogens, they are not relevant to the process of host–pathogen coevolution (Weiss 2002).

7.8 Antibiotic resistance: methicillin-resistant *Staphylococcus aureus*

Pathogens not only evolve in response to the genetic and immunologic resistance of hosts. They also evolve in response to cultural practices that would otherwise reduce their fitness. If the discovery and development of antibiotics was a triumph of modern medicine, the evolution of antibiotic resistance was a perverse triumph of natural selection. The spread of antibiotic resistance among bacteria in the mid-twentieth century was not taken as seriously as it should have been (Dubos 1942). At that time, physicians were not fully aware of the power of natural selection. In addition, new antibiotics were being discovered and made available for clinical use. If a bacterial infection was resistant to penicillin, patients could be treated with streptomycin. When bacteria evolved resistance to streptomycin, patients could be treated with a tetracycline. As long as new antibiotics were in the pipeline, it seemed as though we could stay one step ahead of the bacteria. Now, at least for the time being, we have come to the end of the pipeline, and some bacteria are beginning to evolve resistance to all currently available antibiotics (Levy and Marshall 2004). The emergence of methicillin-resistant

Staphylococcus aureus, or MRSA, is one of the more recent and more worrisome steps in the evolution of antibiotic resistance.

S. aureus is a ubiquitous human companion. Roughly one-half of the population is colonized by this bacterium. Some of us are colonized permanently while others are colonized intermittently. A handful of *S. aureus* strains (or families of related strains) are especially prevalent in humans. These strains have adaptations that increase their colonizing ability. The nose is the most common site of *S. aureus* carriage but the bacteria can also live on other mucous membranes and the skin. Although person-to-person airborne transmission of *S. aureus* can occur, it is thought to be infrequent. The bacteria can survive for long periods outside of the body, and fomites appear to play a major role in their transmission. The bacteria can also be transmitted directly by contact with skin infections.

In the vast majority of people it infects, *S. aureus* behaves as a commensal organism; it is just part of our microbiome. It may stimulate an inflammatory and immunological response but it doesn't cause disease and, by competitive exclusion, the strains we carry may keep us from becoming colonized by more virulent *S. aureus* strains and so may benefit us. But *S. aureus* is an opportunistic pathogen. Occasionally, it causes serious and even life-threatening infections. The bacterium can infect a large number of sites and so can cause a wide variety of clinical infections, including skin and soft tissue infections, pneumonia, systemic sepsis, septic arthritis, osteomyelitis, and toxic shock syndrome. *S. aureus* can harbor many virulence factors and different strains differ in virulence. Although *S. aureus* can be transmitted directly from skin infections, most of these other infections appear to be dead-end infections. The relationship between the virulence and transmissibility of *S. aureus* is not clear.

Shortly after penicillin was introduced into clinical medicine in the 1940s, penicillin-resistant *S. aureus* began to appear. The widespread use of penicillin led to selection for penicillin-resistant bacteria, and the frequency of resistant organisms increased. The gene responsible for penicillin resistance is carried on a plasmid, a circular DNA molecule that can be readily transferred among bacteria but is not integrated into the bacterial chromosome. Recognition that this gene encoded the synthesis of penicillinase, an enzyme that degrades penicillin, led to a search for and development of semisynthetic penicillin derivatives that were resistant to penicillinase. Methicillin, the first of these penicillin derivatives, was made available for clinical use in 1959. By 1961, methicillin-resistant *S. aureus* had arisen.

Penicillin and methicillin kill *S. aureus* by binding to and inhibiting an enzyme involved in the synthesis of the bacterial cell wall. This enzyme was originally identified as a protein that bound penicillin and is now known as penicillin binding protein 2, or PBP2. Methicillin resistance is due to a gene that specifies the synthesis of an altered form of PBP2 that has a lower affinity for methicillin but can still participate in cell wall synthesis. This gene, *mecA*, apparently arose and spread in a different species of staphylococcus and was then transferred into *S. aureus*. Although *S. aureus* is primarily a clonal species (most new alleles arise from mutations rather than from genetic recombination among *S. aureus* strains), it does have several mechanisms to take up and utilize genes from other strains, from other staphylococcal species, and even from unrelated bacteria. One of these mechanisms involves the transfer of what are known as chromosomal cassettes, portions of the bacterial chromosome that can be duplicated in one organism, transferred to another, and integrated into the chromosome of the recipient. The methicillin resistance gene, *mecA*, was transferred into *S. aureus* as part of a chromosomal cassette. Given how rapidly methicillin resistance appeared after the introduction of this antibiotic, it seems unlikely that *mecA* spread initially because it conferred resistance to methicillin. Instead, this gene may have spread over a longer period in response to penicillin, as it would have increased the fitness of staphylococcal species that did not have penicillinase. Then, when methicillin was introduced, the *mecA*-containing cassette may already have been widespread in another species and so was available to be transferred into *S. aureus* (Thurlow et al. 2012).

Methicillin-resistant *S. aureus* strains were originally found in hospital settings, and these organisms became adapted to that environment. Methicillin resistance confers a fitness cost on the bacteria. In the hospital setting, where methicillin is being used, the fitness cost is more than overcome by the fitness advantage of methicillin resistance and so MRSA strains can spread. Because many antibiotics are heavily used in hospitals, most hospital-acquired MRSA strains have evolved resistance to multiple antibiotics. Hospital strains infect sick, often immunodeficient patients, and typically infect open wounds or indwelling catheters. These strains spread on the hands and gloves of health care workers as well as on hospital gowns and equipment (that is, on fomites). Increasing virulence may increase the fitness of these strains, since sick patients will require more frequent and more intensive medical care and so will have greater opportunities to transmit their pathogens to health care workers. Fortunately, the spread of these strains can be controlled by strict adherence to hand washing and other hygienic practices. In other words, good infection-control practices can reduce the R_0 of *S. aureus* to less than one.

Because methicillin resistance imposes a fitness cost on the bacteria, hospital acquired MRSA strains do not spread in the general population. Recently, however, MRSA strains have been found in people outside of hospitals. These so-called community acquired strains are evolving in a different environment than hospital acquired strains and so are evolving different characteristics (David and Daum 2010). The spread of community acquired MRSA strains is a puzzle as well as a serious public health threat. Penicillinase-resistant penicillin derivatives are not widely used outside of hospitals and so the fitness benefits of methicillin resistance in this environment are not obvious. It is not clear why these MRSA strains are spreading. Not surprisingly, MRSA strains are more prevalent in poor and medically underserved communities. Community acquired MRSA strains have other genes that enhance their virulence and colonizing ability. These strains seem preferentially to cause skin and soft tissue infections. The ability to produce skin infections increases the fitness (R_0) of these strains, especially in unhygienic settings where pathogens can be transmitted from these infections either by direct contact or via fomites. The factors that increase transmission by causing purulent skin infections may incidentally increase virulence by promoting infection of other sites. Controlling the spread of community MRSA strains is a major challenge that will require improving the hygiene and overall health of these communities.

7.9 Manifestations of disease

Infectious diseases commonly result in such symptoms as fever, malaise, and pain, and may also be manifested in sneezing, coughing, diarrhea, skin eruptions, etc. These symptoms often lead people with infections to seek medical care. An evolutionary perspective provides a valuable framework for thinking about and understanding these and other manifestations of infectious diseases (Ewald 1980). The signs and symptoms of these diseases may represent host adaptations that decrease the virulence of or fitness loss from the disease, pathogen adaptations that manipulate the physiology of the host in ways that increase the fitness of the pathogen, physiological changes that benefit both the host and the pathogen, or unselected side effects that do not benefit either party. Adaptations frequently involve tradeoffs and so may have harmful as well as beneficial effects. For example, the febrile response is an adaptation that decreases the virulence of infectious diseases but it may also cause febrile convulsions, dehydration, and tissue damage (Kluger et al. 1996). Mammals must have evolved a febrile response to infection because the benefits of fever outweighed its deleterious consequences.

Understanding the pathophysiology that underlies the manifestations of disease can be a useful guide to understanding whether these symptoms are host adaptations or pathogen adaptations. Fever results from the release of cytokines by cells of the native immune system in response to molecules that are recognized as foreign and the action of these cytokines on the temperature regulatory system in the hypothalamus. This understanding supports the recognition that fever is an adaptive response that evolved because (again, on balance) it increased the fitness of host organisms. On the other hand, the diarrhea of cholera is due to a toxin secreted by *V. cholerae* that activates the cystic fibrosis transmembrane conductance regulator we discussed in Chapter 4 and thereby manipulates the normal mechanisms of intestinal secretion. The diarrhea of cholera exemplifies a pathogen adaptation. Diarrhea benefits *V. cholerae* because it flushes much of the microbiome out of the intestinal tract and removes bacteria that would otherwise compete with the pathogen. In addition, diarrhea increases excretion of the pathogen in the feces and its transmission to other people.

Needless to say, decisions to treat patients to relieve the signs and symptoms of disease must be based on appropriate clinical studies and clinical judgment rather than on evolutionary considerations alone. But many "treatments" that were directed at the manifestations of disease—thymus irradiation, for example—have turned out to be harmful, and evolutionary considerations should sensitize us to recognize that the symptoms of disease are not necessarily pathological (Silverman 1993). Evolutionary considerations are also valuable because patients may want to know why they have the symptoms they have, even if this knowledge doesn't affect their therapy. Patients may be reassured to know that their symptoms are part of their bodies' healthy response to an infection.

8

Sexually transmitted diseases

8.1 Introduction

Almost all living organisms have mechanisms for taking up, transferring, or sharing genetic information. We have seen how bacteria can acquire antibiotic resistance and other fitness-enhancing genes from chromosomal cassettes and plasmids. Bacteria have only a single chromosome. When they take up exogenous DNA, this DNA may integrate into or recombine with the bacterial chromosome or it may remain separate. In contrast, most eukaryotic organisms are diploid; they have two copies of each of their chromosomes. These organisms exchange genetic information by reproducing sexually. The life cycle of sexually reproducing species involves the formation of a diploid organism by the combination of two haploid gametes or germ cells, followed by the meiotic recombination of their two genomes in the formation of new gametes.

Sexual reproduction is thought to have arisen early in the history of eukaryotic cells. The reasons why organisms evolved sexual reproduction—that is, the fitness advantages it confers—remain controversial. Two of the leading hypotheses are that sexual reproduction evolved because genetic recombination provides a mechanism to remove deleterious mutations, which would otherwise accumulate in an organism's lineage, or because it increases the genetic diversity of an organism's offspring. Genetic diversity, in turn, may protect these offspring from pathogens or enhance their survival for other reasons, and may thereby increase the reproductive fitness of the parents. Sexual reproduction probably originated with the fusion of two identical haploid cells. Later, this process evolved such that one germ cell, the female or macrogamete, provided almost all of the cytoplasmic constituents of the fertilized zygote, while the other, the male or microgamete, contributed little more than its haploid genome (Maynard Smith 1978).

Given that sexual reproduction may have evolved as a defense against pathogens, it is ironic that it opened up a new route of pathogen transmission. Sexually transmitted pathogens and sexually transmitted diseases (STDs) are widespread in nature. Indeed, sexually transmitted pathogens may have evolved together with the evolution of sexual reproduction. (Only mildly tongue-in-cheek, sperm may be thought of as the first sexually transmitted parasites. They use the nutritional resources of an oocyte to enhance the spread of their own genes. Because their reproduction depends upon the survival and reproduction of the oocytes they infect, sperm have evolved to be benign.) Many plant diseases—diseases known as smuts and rusts—are STDs that are transmitted during pollination. STDs are rare in those fish

Evolution and Medicine. First Edition. Robert L. Perlman © Robert L. Perlman 2013.
Published 2013 by Oxford University Press.

and amphibian species that have external fertilization but are prevalent in species whose reproduction entails close contact between mating partners, and especially in species with internal fertilization, such as birds and mammals (Lockhart et al. 1996).

Humans can be infected by over a dozen pathogens that are primarily transmitted by sexual contact and another several dozen that are occasionally transmitted in this manner (Nahmias and Danielsson 2011). Globally, the incidence of sexually transmitted infections is estimated to be on the order of 1 million/day, or over 300 million new infections per year. Recall that there are roughly 130 million births per year in the world. Despite the uncertainties in these estimates and the transmission of pathogens among men who have sex with men, the probability that heterosexual sexual intercourse will lead to the transmission of a pathogen and to an STD is greater than the probability that it will lead to pregnancy and childbirth. Most STDs are not major killers but HIV/AIDS is thought to cause almost 3 million deaths/year (Lopez and Mathers 2006). HIV is now responsible for more deaths than any other human pathogen.

8.2 The epidemiology of sexually transmitted diseases

STDs differ in interesting ways from other infectious diseases. These differences can be understood in terms of the special features of sexual transmission (Antonovics et al. 2011). When we considered the transmission of other pathogens, we assumed that the entire population was susceptible to and at risk of becoming infected by the pathogen. In the case of sexually transmitted pathogens, only the sexual partners of infected people are at risk of becoming infected. Because infected individuals may have sex many times with a given partner, spread of the pathogen in the population depends on the probability of transmission per partner rather than the probability of transmission per sexual encounter. For sexually transmitted pathogens, then:

R_0 = (number of new sexual partners/time) × (probability of transmission/partner) × (duration of infectivity)

Sexually transmitted pathogens are more likely to be transmitted from healthy people than from sick or dying ones. Infected people must be healthy enough to be sexually active and, leaving aside rape or coerced sex, they must be perceived as healthy and attractive by potential sexual partners. Selection for increased transmissibility, or increased R_0, would therefore be expected to cause these pathogens to evolve to low virulence. Many STDs are characterized by long latent or asymptomatic periods, during which infected people are healthy but infectious. From an evolutionary perspective, the asymptomatic periods of STDs represent evolution of the pathogens to decreased virulence, because infected people might have or raise children and so contribute to their reproductive fitness during this time. From a human perspective, of course, these diseases may still have devastating, albeit delayed, consequences.

In general, then, STDs are chronic diseases. Infected people may remain infected and infectious for long periods of time. For this to be so, pathogens that cause STDs must have evolved mechanisms to evade both our native and adaptive immune responses. Immunity to many sexually transmitted pathogens does not result in permanent resistance. If infected people recover and clear the pathogen from their bodies, they may be susceptible to reinfection. Given realistic estimates of the number of sexual partners that most people have, pathogens that caused acute infections followed by the development of immunity would not be maintained by sexual transmission, because they would not be widely transmitted during the course of the infection and would have R_os less than one.

Perhaps because STDs are chronic diseases with long asymptomatic or latent periods, within-host selection plays an important role in their natural histories. Within-host selection may lead to a high pathogen load and to severe disease or death of infected people. Some sexually transmitted pathogens remain localized in the genital area but some are able to invade other environmental niches in their hosts and so cause systemic disease. Transmission of STDs depends upon pathogens in the genital area rather than on those living elsewhere in the body. Growth of pathogens in these other niches are dead-end infections that do not increase pathogen transmission. Finally, because STDs are chronic diseases and are all spread by the same process, comorbidity is common. Infected people are frequently infected by several sexually transmitted pathogens or by several strains of the same pathogen.

If we think of virulence as a loss of host fitness, we need to consider the effects of the pathogen on fertility as well as on survival. Although infertility may be a very distressing problem to the people who suffer from it, it doesn't carry the toll of premature death—and it doesn't abolish an individual's inclusive fitness, their ability to enhance the survival and reproductive success of their genetic relatives. But infertility of their hosts has very different effects on sexually transmitted pathogens than does the host's death, because these pathogens can be transmitted from infertile people but not from dead ones. Indeed, infertility may well increase the spread of these pathogens, since it may stimulate infertile people or their regular (and possibly infected) sexual partners to seek out new partners. Many STDs do cause decreased fertility. We don't know whether decreasing host fertility is an adaptation that increases pathogen fitness or is simply an unselected side effect of pathogen growth and inflammation in the reproductive tract.

Finally, because sexual contacts are almost always between members of the same species, sexually transmitted pathogens are in general adapted to grow in and be transmitted among hosts of a single species, and they tend to have restricted host ranges. With the exception of HIV, we don't know how sexually transmitted pathogens entered the human population. Like HIV, these pathogens were presumably transmitted from other species to humans by non-sexual pathways and then evolved to be sexually transmitted among humans. Because sexual transmission depends on intimate contacts between sexual partners, sexually transmitted pathogens spend their entire life cycles in or on their hosts and may not have evolved mechanisms that enable them to survive and reproduce outside their hosts. The study of some sexually transmitted pathogens has been hindered by their limited host range and their inability to reproduce outside their hosts. These properties of sexually transmitted pathogens can all be understood as resulting from their mode of transmission.

8.3 Evolutionary responses of hosts to sexually transmitted pathogens

Sexually reproducing organisms face the challenge of maximizing their reproductive success while minimizing the fitness costs of STDs. The evolution of internal fertilization in mammals has been accompanied by the evolution of anti-pathogen defenses in the female reproductive tract. Vaginal and cervical secretions have low pH through much of the menstrual cycle, and they contain inflammatory cells, antibodies, and a number of other proteins and peptides that protect against infection (Profet 1993). The vaginal microbiome is also thought to protect women against sexually transmitted pathogens (Brotman 2011). In addition, the genital area in both men and women is rich in dendritic cells, which play an important role in antigen processing and presentation. The anti-pathogen defenses in the female reproductive tract are similar to those in the skin and respiratory tract. The existence of these defenses is evidence that sexually transmitted pathogens have been important selection factors at some time in—if not throughout—mammalian evolution.

Because the spread of sexually transmitted pathogens depends on the sexual activity of their hosts, these pathogens may have led to the evolution of adaptive changes in host behavior. Most animals limit their sexual activity to times of maximal fertility. Seasonal or periodic breeding can be understood as an evolved life history strategy that conserves resources which can then be used for other purposes, but it may also be an adaptation that limits the spread of sexually transmitted pathogens. The human proclivity to engage in sex throughout adult life, outside of times of peak fertility, may render us especially susceptible to STDs. The monogamous or seasonally monogamous mating systems of some birds and other mammalian species, as well as the limited polygyny of humans, are usually thought of as reproductive arrangements that promote the care and survival of offspring but they too have the additional benefit of minimizing the transmission of STDs and they may have spread for this reason. Cultural norms concerning sexual behavior have played a major role in influencing the spread of sexually transmitted pathogens in human populations.

Because of the great variations in people's sexual activity and behavior, there is great heterogeneity in their risk of acquiring sexually transmitted pathogens. Only sexually active individuals are at risk and the risk increases with the number of sexual partners they have. STDs are not crowd diseases. Their transmission is relatively independent of population density except to the extent that urban lifestyles may differ from rural ones and city dwellers may have more sexual partners than people who live in the country. Epidemiologists usually model the spread of sexually transmitted pathogens in terms of a "core population" with a high rate of acquisition of new sexual partners and a high probability of acquiring STDs, and a larger population with a lower rate of partner acquisition and a lower risk of disease. These core high-risk populations may be sex workers and their clients or other groups who have multiple sexual partners. R_0 within the core group is greater than one, while R_0 outside of the core is less than one. Pathogens are maintained in the population as a whole because of their spread within the core group and their occasional transfer from people in the core to the rest of the population.

Sexually transmitted pathogens may also be transmitted "vertically," from parents to their offspring. Although other pathogens can also be transmitted from parents to offspring, sexually transmitted pathogens are especially likely to be transmitted in this way. These pathogens may have access to the germline of the fetus and so may become incorporated into their host genome. Some sexually transmitted pathogens "hitchhike" through the female reproductive tract by binding to sperm. It is easy to imagine that these organisms might occasionally be taken up into fertilized zygotes along with the chromosomes from the sperm. The human genome contains many endogenous retroviruses—or, more accurately, retrovirus-like sequences that are inactive and are not transmissible. These endogenous retroviral sequences, which may comprise 1–2% of the human genome, are most likely the residue of sexually transmitted retroviruses whose DNA retrotranscripts became integrated into human germline DNA but then lost the ability to replicate. In addition, the adaptations that enable sexually transmitted pathogens to evade host immune defenses may also enable them to cross the placenta. *Treponema pallidum*, the bacterium that causes syphilis, is one of the few bacteria that can cross the human placenta and cause disease in the fetus. Finally, sexually transmitted pathogens in the genital area may be transmitted to newborns at the time of delivery. Several sexually transmitted pathogens, including most importantly *Neisseria gonorrhoeae*, can cause conjunctivitis and blindness in newborns. Pathogens that evolve to be transmitted exclusively by vertical transmission would be expected to evolve to be benign, i.e. not to decrease the fitness of their hosts. This is presumably the evolutionary history of endogenous retroviruses. Most vertically transmitted pathogens, including *T. pallidum* and HIV, are virulent in the offspring they infect and are not transmitted efficiently from these infected offspring. Vertical transmission of these organisms is another example of dead-end evolution.

8.4 Syphilis

Syphilis exemplifies the evolution and natural history of STDs (Quétel 1990). Syphilis appeared in Europe as a new disease that erupted in 1495 in the army of Charles VIII of France at the battle of Fornovo, when the army was retreating from Naples. Historical accounts suggest that it probably existed in Spain in 1494; it may have been brought to Italy by Spanish soldiers or camp followers. The initial rapid spread of syphilis was fueled by the high number of sexual contacts between soldiers and prostitutes, which in turn was fueled by the social dislocations of war. Early accounts describe syphilis as a virulent, often rapidly fatal disease. The disease was known originally as the Great Pox, because victims were covered with large skin eruptions, or pox. (The other fatal disease that was manifested by skin eruptions then became known as the Small Pox.) During the sixteenth century, syphilis became much less virulent and came to resemble the disease as we know it today. The decreased virulence of syphilis may have been due in part to increased resistance of the human population, those most susceptible to the disease having died of it during the first hundred years after it emerged, but was probably due primarily to selection for increased transmissibility of *T. pallidum*, the bacterium that causes it. While syphilis is no longer the scourge it used to be, it is still an important disease, both in the United States and in developing countries. The incidence of syphilis in the United States is now on the order of 10 thousand new cases per year; globally, there are over 10 million new cases per year. In the United States, syphilis is primarily a disease of the urban and rural poor, and of intravenous drug users, but in other countries it is more widespread.

Evolution of *Treponema pallidum*

Syphilis is caused by a subspecies of *T. pallidum* known as *T. pallidum* pallidum. *T. p.* pallidum is apparently a young subspecies, as shown by the very few polymorphisms found in those strains whose genomes have been sequenced. The origin of *T. p.* pallidum has long been a subject of controversy. Although the timing of the appearance of syphilis in Europe is consistent with the idea that *T. p.* pallidum was brought back from North America by Columbus's sailors, there is only scant and inconclusive evidence for the existence of syphilis in the New World at that time. *T. p.* pallidum is closely related to several other treponemas, most notably *T. p.* pertenue, the organism that causes yaws. Yaws, a skin infection that is transmitted by direct skin-to-skin contact, is common in tropical environments in both the New World and the Old. Yaws is thought to be an older disease than syphilis, and *T. p.* pertenue appears to be older than *T. p.* pallidum. *T. p.* pallidum apparently arose from a strain of *T. p.* pertenue that evolved to be transmitted sexually. Direct skin-to-skin transmission is similar to sexual transmission in that both routes require intimate touching and so transmission is likely to be among members of the same species. Indeed, directly transmitted skin microorganisms may be transmitted during the intimate contact of sexual intercourse. The biggest difference is that while skin-to-skin transmission is common among children, sexual transmission is almost exclusively limited to adults.

T. p. pallidum and *T. p.* pertenue are so similar morphologically, immunologically, and genetically that until recently they could only be distinguished by the diseases they caused. Even now, although a handful of genetic differences between these bacteria have been reported, only a few strains of each pathogen have been sequenced and these genetic differences do not obviously account for the different modes of transmission of the pathogens. If *T. p.* pallidum did evolve from *T. p.* pertenue, only a small genetic change was required to change the route of pathogen transmission.

Although limited sequencing studies have suggested that *T. p.* pallidum is most closely related to New World strains of *T. p.* pertenue, more complete sequencing of more strains of both *T. pallidum* subspecies will be required to determine the origin of *T. p.* pallidum definitively (Harper et al. 2008; Mulligan et al. 2008). Fortunately, the progress being made in the eradication of yaws may preclude such studies. Even if *T. p.* pallidum arose from a New World strain of *T. p.* pertenue, it isn't clear whether the transition from skin-to-skin transmission to sexual transmission occurred before Columbus and his sailors came to the New World or after they returned to Europe. A sexually transmitted pathogen could only have spread in an environment in which the R_0 for sexual transmission was greater than one. While this was evidently the situation in Europe at that time, it may not have been the case in the New World.

One hypothesis, which is now out of favor but is still attractive, proposes that syphilis and yaws are due to the same pathogen and that the differences between the diseases are due to ecology rather than genetics. According to this "Unitarian" hypothesis, in hot, damp environments where skin is often uncovered and skin lesions are oozing, *T. pallidum* will be readily transmitted among children and cause yaws. These children develop immunity to yaws that also protects them from acquiring syphilis. In colder climates where skin is covered, children do not become exposed to the pathogen until they become sexually active. If they then acquire *T. pallidum* sexually, they develop syphilis (Hudson 1965). Our inability to culture *T. pallidum* in vitro and our limited ability to grow the bacteria in other species make it difficult to resolve the controversies over the origin of *T. p.* pallidum and its relationship to *T. p.* pertenue.

Natural history of syphilis

Syphilis has a complicated and varied natural history (LaFond and Lukehart 2006). Primary syphilis is typically manifested by an ulcerating lesion, or chancre, on the skin or mucous membranes, which appears several weeks after infection. Chancres usually last for several weeks and almost always heal without treatment. Although the symptoms of primary syphilis are localized to the site of infection, *T. pallidum* spreads to many other tissues during this time. The widespread dissemination of the bacteria throughout the skin and mucous membranes leads to generalized mucocutaneous rashes, which typically appear shortly after chancres heal and constitute what is known as secondary syphilis. The symptoms of secondary syphilis may last for a year or more but again usually resolve without treatment. The symptoms of primary and secondary syphilis are relatively mild and may even go unnoticed but infected people remain infectious throughout the periods of primary and secondary syphilis. After this time, the infection may be cleared by the immune system or it may become latent. Many (typically 20–30) years later, some people with latent infections develop tertiary syphilis, with growth of bacteria in the central nervous system and/or the cardiovascular system, bone, and other tissues. Until the advent of penicillin, tertiary syphilis was a major cause of disability and death. Tertiary syphilis results from the renewed growth of bacteria that spread to these other tissues early in infection but remained quiescent. The recrudescence of *T. pallidum* is not understood but is more likely to be the result of the waning of host defenses than of within-host selection for mutant organisms that can grow in these niches. People with tertiary syphilis are usually no longer infectious. In any event, bacteria growing in the brain or heart would not be transmitted to sexual partners. Tertiary syphilis represents dead-end infections of these tissues.

    T. pallidum has evolved several strategies that enable it to evade our immune defenses and to cause long-lasting infections. It has a dearth of integral membrane proteins and of cell surface antigens. One of these antigens, a protein known as TprK (*T. pallidum* repeat protein K) has several variable regions, and the bacteria have evolved a mechanism to recombine a large number of alternative sequences into these variable regions. An immune response to one of these variants leads to selection for other variants. Because of the large number of possible variants, this process can be repeated over and over during the course of an infection (Giacani et al. 2010). *Tprk*, the gene that encodes TprK, is the only gene that is known to undergo selection during the course of a syphilis infection.

The paucity of *T. pallidum* cell surface proteins has several consequences. The lack of nutrient transport proteins is one reason why *T. pallidum* grows much more slowly than other bacteria. Also, *T. pallidum* appears not to have cell surface receptors for viruses or plasmids, and so it has no mechanisms to take up, exchange, or transfer DNA. The benefits of immune evasion evidently outweigh the deleterious consequences of the absence of membrane proteins. But the apparent inability of *T. pallidum* to take up and express exogenous DNA is a puzzle, because the bacteria have no known mechanism to correct mutations. In any event, this inability has an important clinical benefit: *T. pallidum* is one of the few bacteria that have not evolved resistance to penicillin.

Cultural responses to syphilis

Syphilis has not only been a major cause of illness and death over the last 500 years, but has arguably had a greater impact on Western culture than any other disease. If syphilis wasn't the source of our sexual norms, the value our culture has placed on chastity and monogamy, and our antipathy to prostitution, it certainly reinforced these beliefs. Syphilis was recognized to be a sexually transmitted disease almost from the beginning, but at that time there was no such thing as "safe sex" and syphilis could be prevented only by monogamy or sexual abstinence. Condoms were not commercialized until the beginning of the eighteenth century. Although they were certainly used for contraception, they received societal acceptance for the prevention of syphilis. Most importantly, the fear of syphilis disrupted the fabric of society and eroded the bonds of fellowship and trust that prevailed before the onslaught of the disease (Crosby 1971, p. 158). In our time, HIV has evoked similar cultural responses, including changed patterns of sexual behavior, increased condom use, and, at least initially, fear of contact with and ostracism of people who were suspected of being infected with the virus. Sexually transmitted diseases, like other infectious diseases, elicit the disgust response we discussed in Chapter 7. Perhaps because sex itself is such an emotionally fraught subject in our culture, this response can easily be transformed into unwarranted hostility toward and victimization of people who are suffering from STDs.

8.5 HIV/AIDS

Human immunodeficiency virus (HIV) causes the disease Acquired ImmunoDeficiency Syndrome (AIDS). (There are two human immunodeficiency viruses, HIV-1 and HIV-2. Because HIV-1 is by far the more important human pathogen, we will focus our attention on this virus.) AIDS was recognized as a disease in 1981 and HIV was isolated in 1983. HIV is still spreading so rapidly and new therapies are changing the natural history of HIV infections so dramatically that epidemiological statistics rapidly become dated, but some recent estimates from the U.N. Program on HIV/AIDS (UNAIDS) indicate the magnitude of the burden of human disease caused by HIV. Some 30–35 million people are thought to be infected with HIV and another 2.5–3 million people per year are becoming newly infected. Of the 30–35 million infected individuals, approximately two-thirds live in sub-Saharan Africa. Up to 30% of adults in some urban areas of sub-Saharan Africa carry the virus. India, with 3–5 million HIV-infected people, is the country with the greatest prevalence, but this is less than 1% of the Indian population. In the 30 years since AIDS was identified, it has been responsible for some 25–30 million deaths and, as mentioned earlier, it is currently causing close to 3 million deaths per year.

Although HIV is commonly thought of as a disease of men who have sex with men and of intravenous drug users, globally upwards of 75% of HIV infections result from heterosexual sex, 5–10% from sex between men, and the remainder from shared needles and syringes, and from transfusion with contaminated blood or blood products. HIV can also be transmitted from infected pregnant women to their children, either in utero or in the neonatal period. Most children who acquire HIV from their mothers, either prenatally or postnatally, develop AIDS in childhood. Until now, at least, these infected children have not been an important source for onward transmission of the virus.

Evolution of HIV

Viruses similar to HIV, called simian immunodeficiency viruses (SIVs), are widespread among monkeys. SIVs apparently can cross species barriers easily and have now infected several dozen species of monkeys. HIV (specifically, HIV-1) viruses are more similar to chimpanzee SIV than to other SIVs and have almost certainly evolved from the chimpanzee virus, which is known technically as SIV_{cpz} (Sharp and Hahn 2011). SIV_{cpz} appears to be sexually transmitted among chimpanzees and to cause an AIDS-like illness in at least some of the animals it infects. The existence of four genetically distinct groups of HIV-1 strains, each of which is more similar to chimpanzee SIV than to the other HIV strains, suggests that SIV_{cpz} has been transmitted from chimpanzees to humans on multiple occasions. One of the four HIV-1 groups, group M, is responsible for the global epidemic of HIV; until now, the other groups have been of only minor importance. The ancestor of group M viruses apparently entered the human population in the southeast corner of Cameroon. It was most likely transmitted from a chimpanzee to a hunter, either from a bite or from exposure to blood or other body fluids.

The genetic diversity of present-day group M strains suggests that the ancestors of this virus began spreading some time in the first third of the twentieth century. The early spread of the virus is thought to have occurred in Léopoldville, which is now Kinshasa in the Democratic Republic of the Congo. The first documented case of HIV was in Kinshasa and the greatest diversity of group M viruses is found in that region. How the virus traveled from the first infected person in the remote, forested location of southeast Cameroon to the urban area of Léopoldville is not known for sure but is very likely to have involved a stage of inadvertent "parenteral amplification" of the virus by medical workers. At that time, public health programs were treating patients suffering from sleeping sickness, leprosy, yaws, or syphilis with intravenous or intramuscular drug injections. Because of limited resources and limited appreciation of the risks involved, the teams that carried out these treatments reused syringes and needles without adequate sterilization. Spread of the virus via contaminated needles may have greatly increased the population of HIV-infected people and made it more likely that one such person would somehow get to Léopoldville. There was a window of about 50 years between the launching of these public health programs and the institution of strict sterilization procedures. It is probably not a coincidence that HIV gained a foothold in the human population during this period (Pepin 2011).

At the time HIV is thought to have reached Léopoldville, the city had a large excess of single men and an active prostitution industry, conditions that would have enabled the virus to flourish as a sexually transmitted pathogen. Nonetheless, it seems to have taken at least 50 years between the initial spread of the virus in Central Africa and the time it erupted into a global epidemic. We don't know anything about the natural history of HIV infections until the first reports of this disease in the 1980s and so we don't know how HIV evolved during this period. It seems likely that part of the delay may have been the time necessary for HIV to evolve from an injection-transmitted to a sexually transmitted pathogen in humans. Since its initial description in the 1980s, HIV has apparently evolved to be more virulent, as measured by a more rapid decline in $CD4^+$ cells and a higher level of virus during the latency period (Muller et al. 2009). Interpreting the change in virulence is complicated, since it seems to be more pronounced among drug addicts than among people who acquire HIV sexually. Transmission by injection might be expected to select for increased virulence. The multiple routes of HIV transmission complicate the relationships between virulence and transmissibility of the virus.

Natural history of HIV infections

When HIV is sexually transmitted to a new host, it begins to replicate in lymph nodes in the genital area and the gastrointestinal tract. There is typically a latent period of several weeks after infection before the virus appears in the blood and then the viral titer in blood increases, peaking about 6 weeks after infection. Even in the absence of therapy, virus levels then decline and the disease enters another latent period, which may last from several months to more than 15 years. During this latent period, the virus is hardly quiescent; perhaps on the order of 10^{10} virus particles and 10^7 virus-infected cells are produced per day. Finally, virus levels increase again and the symptoms of AIDS (immunodeficiency, infections with other pathogens) appear. Death usually occurs within 2–3 years after the onset of AIDS (Forsman and Weiss 2008). People who are infected with HIV are infectious throughout the course of their illness. During the lifetime of an infected person, most sexual transmission occurs during the latent, asymptomatic period. Although virus titers are much lower during this period than either early in infection or at the onset of AIDS, the asymptomatic period lasts for a much longer time (Hollingsworth et al. 2008).

HIV is virulent because it infects and kills cells of the immune system and causes immunodeficiency. It infects several cell types, including T lymphocytes, macrophages, and dendritic cells. In particular, the virus infects a subset of T cells known as $CD4^+$ cells, which express the CD4 protein on their cell surface. The progress of the disease is typically monitored by measuring the level of $CD4^+$ cells in blood. $CD4^+$ cells fall during the initial infection, rebound to normal levels (about 1000/mm³), and then gradually decline during the latent period. A fall to about 200 cells/mm³ signals or accompanies the onset of AIDS.

HIV specifically infects CD4⁺ cells because a protein on the viral envelope known as gp120 specifically binds CD4. CD4 appears to be the major cellular receptor for HIV; it is necessary but not sufficient for viral entry into cells. Another T-cell protein, CCR5, appears to be an especially important co-receptor. Early in sexually acquired infections (but not in infections acquired by injection), the virus is restricted to the subset of CD4⁺ cells that also express CCR5. In the course of an infection, there is within-host selection for viruses that can utilize other proteins as co-receptors and so can infect other subsets of CD4⁺ cells. Infection of a larger fraction of T cells helps to explain how the virulence of HIV increases during the course of an infection.

CD4 is present on T cells, on macrophages, and on dendritic cells. Although the course of HIV infections may be monitored by measuring the level of CD4⁺ cells in blood and the loss of CD4⁺ cells is probably responsible for immunodeficiency and AIDS, transmission of the virus and infection of a new host depends upon infection of macrophages and dendritic cells. The requirements for infecting macrophages and dendritic cells, and for binding to CD4 and CCR5, presumably result in stabilizing selection during transmission of the virus and so reduce variation among the viruses that are circulating in a community.

HIV is a retrovirus. It carries its genetic information in the form of RNA rather than DNA. When the virus enters a host cell, the enzyme reverse transcriptase makes a retrotranscript, a DNA sequence that is complementary to the viral RNA. The viral DNA is inserted into the genome of the host cell and is then transcribed to produce new viral RNAs. Translation of the viral RNA produces reverse transcriptase and the other proteins that are found in the mature virus. One reason for the persistence and the virulence of HIV is its high mutation rate. Reverse transcriptases have a high error rate and retroviruses do not have a mechanism to repair mistakes in transcription. The error rate for HIV DNA synthesis is on the order of 10^{-4}/nucleotide/cycle. Given that the HIV genome contains about 9000 nucleotides, there is roughly one mutation in each cycle of viral replication. And given the rate at which viral particles are produced, even during the asymptomatic period, every possible HIV mutation, and many combinations of mutations, are probably occurring daily (Richman et al. 2004). Genetic recombination is also important in the generation of HIV diversity. The mature, infectious virus contains two strands of RNA. It is not a double-stranded RNA virus; both of the viral RNA strands encode viral proteins. Because of the high mutation rate and because people may become superinfected with several HIV strains, the two RNA strands in a virus may not be identical. When the virus enters a host cell, reverse transcriptase begins to synthesize viral DNA while the viral RNAs are still held together within the viral capsid. Reverse transcriptase may jump from one viral template to the other during DNA synthesis, creating recombinant viral DNAs. Because of genetic recombination as well as its high mutation rate, the HIV population in an infected person shows great variation. This variation provides abundant raw material for the selection of viral mutants—selection to evade host immune responses, to infect new cell types during the course of an infection, and to evolve resistance to anti-viral drugs. Within-host selection is diversifying selection. The diversity of viruses within an individual infected person is greater than the diversity of viruses that are being transmitted in the population.

Human responses to HIV

There is great variation in the natural history of HIV infections. While some of this variation is due to differences among viral strains, much appears to be due to differences among people. Some people have a natural resistance to HIV. Even after repeated exposure to the virus, they do not have detectable levels of virus or of anti-viral antibodies. Others do become infected but maintain high levels of $CD4^+$ cells and low levels of virus, and remain healthy for many years after becoming infected, even in the absence of therapy (Poropatich and Sullivan 2011). The best-known mutation that is associated with resistance to HIV is a variant CCR5 allele, CCR5Δ32, which carries a 32 base pair deletion; the protein encoded by this allele does not bind gp120. People who are homozygous for the CCR5Δ32 allele are resistant to sexually acquired HIV, but not to HIV acquired by injection; heterozygotes may show a prolonged latent period between infection and the onset of AIDS. Roughly 10% of the Caucasian population carries the CCR5Δ32 allele. It is found in highest frequency in Northern Europe and Western Asia. Although CCR5Δ32 may have spread because it conferred resistance to some earlier pathogen, it did not spread as an adaptation to HIV. There is as yet no clear explanation for the prevalence and geographic distribution of this allele. As we discussed earlier, the spread of HIV must be causing selection for resistance to the virus but there have not been measurable genetic changes in humans during the short time that we have been living with the virus.

HIV therapy

The first drugs that were used to treat people with HIV infections were reverse transcriptase inhibitors. The use of single drugs predictably resulted in selection of drug-resistant viruses. The failure of single-drug therapy led to the development of combination therapies for HIV. The development of combination therapies was motivated at least in part by evolutionary considerations. The use of two different types of reverse transcriptase inhibitors, which inhibited the enzyme by different mechanisms, was based on the premise that resistant mutants would require at least two mutations in reverse transcriptase and would likely have decreased fitness (Chow et al. 1993). The development of combination anti-retroviral therapy has had a dramatic impact on the natural history of HIV infections. Drug therapy controls the disease in most patients and has turned HIV from a progressive, lethal disease into a chronic, manageable disease. Drug treatment is not without its problems. It is expensive, it depends on patients actually taking their medication, and it has side effects. But in addition to prolonging the healthy lives of people who are infected with HIV, drug therapy appears dramatically to reduce the risk of transmission of the virus. Combination drug therapy has created new optimism that we may finally be able to curb the spread of HIV and reduce the toll from AIDS.

9

Malaria

9.1 Introduction

Most people in the United States and other economically developed countries are comfortably isolated from malaria. Even when we travel to countries where malaria is endemic and we have to take prophylactic medication, it may be difficult for us to appreciate the global impact of malaria on morbidity and mortality. Despite recent progress in reducing the burden of malaria, there are still over 200 million new cases of malaria a year and well over a million deaths a year from the disease, more than half of which are in infants and young children (Murray et al. 2012). Malaria has probably been a major cause of infant and childhood deaths for millennia. Without question, it is the infectious disease that has had the greatest evolutionary impact on human beings and that most clearly illustrates the principles of human-pathogen coevolution (Kwiatkowski 2005).

Not only is malaria important, it is also complicated. Malaria is caused by parasites in the genus *Plasmodium*. Four plasmodium species—*P. falciparum*, *P. malariae*, *P. ovale*, and *P. vivax*—have classically been recognized as causes of malaria in humans. Recently, a fifth species, *P. knowlesi*, has been identified as an emerging human pathogen. Each of these species has its own evolutionary history and specialized adaptations. Moreover, these parasites may be transmitted to humans by dozens of species of mosquitoes, again with their own unique ecologies and behaviors. Because *P. falciparum* is by far the most virulent and medically important human malaria parasite, our discussion will focus on this species.

9.2 The life history of *Plasmodium falciparum*

Plasmodia are single-celled eukaryotic organisms. Their life cycle comprises asexual reproduction in a vertebrate host and sexual reproduction in an arthropod (mosquito) vector. There are over 200 species of plasmodia, which infect a variety of reptile, bird, and mammalian hosts and which are transmitted by a variety of arthropod vectors. Although there are differences in details, the life cycle of *P. falciparum* may be taken as representative of that of other plasmodium species (Figure 9.1). Infection begins when a susceptible person is bitten by an infectious mosquito. Plasmodia in the stage of their life cycle known as sporozoites are present in the salivary glands of infected female mosquitoes (only females eat blood meals and so only females are infectious) and are injected into the victim at the time of the bite.

Evolution and Medicine. First Edition. Robert L. Perlman © Robert L. Perlman 2013.
Published 2013 by Oxford University Press.

The sporozoites circulate in the blood until they reach the liver, where they enter hepatocytes and perhaps other liver cells. Sporozoites are rapidly cleared from the blood; they enter
liver cells within a few hours after a bite. In the liver, sporozoites undergo several rounds of
asexual replication and differentiate into a new form, known as merozoites. Fewer than 100
sporozoites are delivered into the blood in an infectious bite but these sporozoites may give
rise to tens of thousands of merozoites. After merozoites are released from the liver, they
enter red blood cells and replicate, again asexually, in the red cells. A round of replication of
P. falciparum merozoites takes 48 hours and results in the lysis of erythrocytes and the
release of more merozoites, which then invade other red cells and begin a new round of replication. On average about 16 merozoites are released from each infected erythrocyte but
there is considerable variation in this number. In severe infections, up to 50% of an infected
person's red cells may be infected and parasite densities may reach tens of thousands per
cubic millimeter (microliter) of blood. An infected person may be home to billions of parasites. While most of the merozoites that enter red blood cells replicate and produce more
merozoites, some differentiate into progenitors of germ cells, known as gametocytes. There
are two types of gametocytes, macro (female) and micro (male). Gametocytes remain within

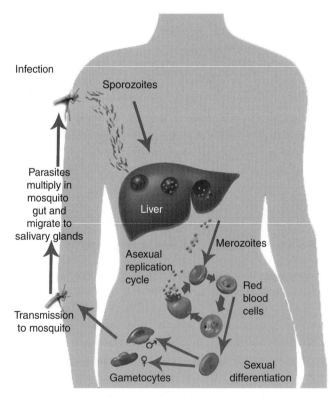

Figure 9.1 Life cycle of *Plasmodium falciparum*.
Modified from <http://history.nih.gov/exhibits/bowman/SSmalaria.htm>

the erythrocytes but do not replicate there. Although gametocyte-infected red cells are phagocytosed and their gametocytes are degraded, new gametocytes continue to be produced and can accumulate to densities of hundreds or thousands per microliter over the course of an infection.

If a mosquito bites an infected person, she may consume some gametocytes as part of her blood meal. In the mosquito, the gametocyte-containing red cells lyse and the gametocytes differentiate into macrogametes (oocytes) or microgametes (spermatocytes). Plasmodia are haploid throughout the portion of their life cycle that they grow in humans (i.e., as sporozoites, merozoites, and gametocytes). Fusion of a macro- and microgamete produces a diploid zygote, or ookinete. The ookinete migrates through the intestinal wall of the mosquito, where it grows, undergoes meiosis, and differentiates to produce sporozoites, which migrate to the mosquito's salivary glands and become competent to infect a new victim.

Because macrogametocytes and microgametocytes are produced by differentiation of haploid merozoites, they (and the gametes that arise from them) are genetically identical. Sex of the gametocytes and gametes must be determined by epigenetic or environmental factors. *P. falciparum* undergoes a significant amount of inbreeding or self-fertilization by genetically identical gametes. Self-fertilization may be beneficial because it removes deleterious mutations but it would not increase genetic diversity among the resulting sporozoites. In areas where malaria is prevalent, mixed infections or superinfections—infection by two or more *P. falciparum* strains—are common. Cross-fertilization can occur when a mosquito bites a person who is infected by multiple *P. falciparum* strains and whose blood harbors gametocytes with different genotypes or, less commonly, when a mosquito bites two infected people in relatively rapid succession. Cross-fertilization would increase genetic diversity of the offshoot plasmodium population.

The production of macrogametocytes is usually greater than the production of microgametocytes, typically on the order of three or four to one. While each macrogametocyte produces only one macrogamete, each microgametocyte gives rise to up to eight microgametes, so that the sex ratio of gametes in the mosquito is closer to one to one. But the parasites can alter the sex ratio of their gametocytes. Mixed infections usually lead to a greater production of microgametocytes, presumably because the resultant microgametes have to compete for access to macrogametes (Talman et al. 2004).

When merozoites invade red blood cells, they must balance the production of more merozoites with the production of gametocytes. When viewed as life history strategies, production of merozoites represents investment in growth and maintenance of the parasite population (and keeps open the possibility of future reproduction) while production of gametocytes represents utilization of resources for more immediate reproduction. Malaria parasites have evidently evolved mechanisms that sense the quality of the host environment and regulate gametocyte production in ways that optimize their reproductive fitness. When the host environment is favorable for growth of the parasite population, *P. falciparum* invests more resources in merozoite production and produces relatively few gametocytes. On the other hand, when immune responses or drug treatment renders the host environment unfavorable and parasite densities are decreasing, the parasite devotes more of its energy to producing gametocytes. Even when survival of the parasite population is in jeopardy, only a small percentage of *P. falciparum* merozoites differentiate into gametocytes. Natural selection may have limited gametocyte production because too-high levels may elicit immune responses that inhibit transmission or because the ingestion of too many gametocytes decreases the viability of mosquitoes (Talman et al. 2004).

9.3 The natural history of malaria infections

The natural history of malaria infections is closely related to the life cycle of *P. falciparum*. Sporozoites may kill the handful of liver cells they infect but they don't cause any detectable illness. The symptoms and pathophysiology of malaria are due, directly or indirectly, to the growth of merozoites in red blood cells. Because a round of *P. falciparum* merozoite replication takes 48 hours and the release of merozoites from red cells is accompanied by fever, *P. falciparum* malaria causes a characteristic pattern of fever. At least early in infections, patients get spikes of fever every 48 hours, typically in the evening, following a round of red cell lysis and release of merozoites. Some substances that are released upon lysis of plasmodia-infected red blood cells stimulate macrophages to release tumor necrosis factor-α and other inflammatory cytokines, which then produce fever and other systemic manifestations of inflammation (Richards 1997). Later in the course of infection, the fever spikes may become dampened, as different broods of merozoites develop asynchronously. Although, as we discussed earlier, the febrile response to infection is on balance an adaptation that benefits people who are suffering from infectious diseases (Kluger et al. 1996), the unusual, stereotypical febrile response to malaria is evidently a manipulation of our physiology by the parasite. The periodic fever caused by *P. falciparum* is likely to enhance parasite transmission by making infectious people more attractive to mosquitoes and debilitating these people at just the time of day that mosquitoes are active and biting. The synchronized maturation of merozoites and their release from red cells appears to be regulated by the hormone melatonin, which has a circadian rhythm of concentration and modulates the cell cycle of *P. falciparum* (Beraldo and Garcia 2005). The response of malaria parasites to melatonin is another way in which these parasites have evolved to take advantage of our physiological processes.

As merozoites replicate in red blood cells, they produce several proteases that degrade hemoglobin, which provides a major source of nutrition for the parasites. The loss of hemoglobin, together with increased clearance of infected red cells in the spleen, increased intravascular hemolysis of uninfected red cells, and reduced erythropoiesis, results in anemia, which is one of the important causes of disability and death in patients with malaria. In addition, merozoites express a number of proteins that are inserted into the red cell membrane. Some of these proteins cause infected erythrocytes to aggregate and to bind to endothelial cells in capillaries and venules. Adherence to endothelial cells appears to be an adaptation that decreases the exposure of parasites to the immune system and shields them from destruction in the spleen. But the aggregation of infected red cells in capillaries and venules may reduce blood flow to critical organs and when these infected red cells rupture, the local release of cytokines and other substances may further disrupt the function of these organs.

The aggregation of plasmodia-infected red cells in cerebral vessels may lead to cerebral malaria, which is manifested by convulsions and coma, and which is another important cause of disability and death in malaria. Aggregation of infected red cells in the renal circulation may lead to tubular necrosis and renal failure. And the same pathological process in the placentas of pregnant women may result in stillbirths or decreased fetal growth. Although cerebral malaria probably increases parasite transmission because it debilitates infected people and makes them more susceptible to insect bites, the other complications of *P. falciparum* infections do not appear to enhance parasite fitness. These complications are presumably unselected side effects of the aggregation of infected erythrocytes in the microcirculation.

The first symptoms of malaria typically appear about ten days to two weeks after a person has been bitten by an infected mosquito. It takes about a week for merozoites to be released from the liver and then another two days for the first round of merozoite replication in and rupture of red blood cells, but the first few rounds of merozoite release may be asymptomatic. Malaria infections can last for a very long time, perhaps on the order of ten months or a year. Ultimately, parasites are cleared from the blood and infected people develop some adaptive immunity to *P. falciparum*. Immunity to malaria is good but far from perfect. People who survive *P. falciparum* infections can become reinfected multiple times, but subsequent infections are usually milder than the initial episode. In areas where malaria is endemic, it is predominantly a disease of young children. Because they have longer infections and higher gametocyte levels than do adults, infants and children play an especially important role in malaria transmission. Older people, who have survived multiple attacks of malaria, have some acquired immunity that limits the virulence of new infections. Nonetheless, malaria remains an important if unappreciated cause of death in adults (Murray et al. 2012).

P. falciparum and other malaria parasites have evolved a fascinating mechanism to evade host immune responses and produce chronic infections. The parasites spend most of their life cycle sequestered in red blood cells, where they are not directly exposed to the immune system. Parasite proteins that are inserted into the red blood cell membrane can evoke an immune response. Many of these antigenic parasite proteins are encoded by gene families, and parasite populations can switch expression of these genes so that they display new antigens over the course of an infection. One of the parasite proteins that participates in the binding of infected red cells to endothelial cells is a protein known as PfEMP1 (*P. falciparum* erythrocyte membrane protein 1). PfEMP1 is a product of the *var* or variable genes. The *P. falciparum* genome contains on the order of 60 *var* genes. The parasites express only one var gene at a time but they can change the expression of their *var* genes periodically. Expression of *var* genes is apparently under epigenetic control, since the expression of an individual *var* gene (and silencing of the others) is maintained over many rounds of cell division. As the immune system recognizes and responds to one PfEMP1 protein and attacks cells that express this protein, variant parasites that express other, antigenically distinct PfEMP1 proteins are selected and spread in the infected host (Scherf et al. 2008). Malaria infections are often characterized by waves of parasitemia, each wave being due to the spread of parasites expressing a new PfEMP1 protein. Although the mechanism of antigenic variation in *P. falciparum* differs from that in *T. pallidum*, it has the same fitness benefit of enabling the pathogen to evade the immune responses of the host and to cause long-lasting infections.

People who are infected by *P. falciparum* become infectious about two weeks after the onset of the infection and remain infectious for the duration of their infection. Again, it takes about a week before merozoites are released from the liver and another week or more for the first gametocytes to mature in infected red blood cells. Although gametocytes accumulate in the blood over the course of an infection, they do not appear to cause symptoms or to injure their hosts.

9.4 R_0, the basic reproductive number of *P. falciparum*

Although *P. falciparum* can cause epidemics, most parasite transmission occurs in regions where it is endemic. As with other pathogens, we can consider the basic reproductive number, R_0, to be a good measure of *P. falciparum* fitness and we can expect natural selection to optimize the R_0 of the parasite. Because of the complex life cycle of malaria parasites, the R_0 of *P. falciparum* is more complicated than that of directly transmitted pathogens. The number of secondary *P. falciparum* infections that would result from the introduction of one infectious person into a population of susceptible individuals equals the number of mosquitoes that would on average become infected by biting an infected person times the number of people who would on average become infected by a single infected mosquito.

Transmission of malaria from humans to mosquitoes depends on the duration of infectivity in people, the bite rate (the average number of mosquito bites an infected person gets per day), and the probability that a bite will result in transmission of the parasite from an infected person to a mosquito. Likewise, transmission from mosquitoes to humans will depend on the duration of infectivity in mosquitoes, the bite rate (here, the average number of human bites an infectious mosquito makes per day), and the probability that the bite of an infectious mosquito will lead to an infection in the bitten person. Even though we can't quantify all of the factors that contribute to R_0, this framework provides a basis for thinking about the evolution of *P. falciparum*. Smith et al. (2007) provide a more rigorous discussion of the R_0 of malaria. Most importantly, R_0 is not a fixed characteristic of the parasite, because it depends on many factors that the parasite can't control. Because the bite rate plays a critical role in the transmission of malaria from people to mosquitoes and from mosquitoes to people, R_0 depends heavily on the density and behavior of the local mosquito population, as well as on the life expectancy of the mosquitoes, the density of the host population, and the availability of other species for the mosquitoes to feed upon. Estimates of R_0 for *P. falciparum* in 121 African populations in different environments range from around one to over 3000 (Smith et al. 2007).

P. falciparum infections typically last many months, or hundreds of days—this long infectious period increases the opportunity for transmission of the parasites to mosquitoes. Individual mosquitoes may make on the order of 1–10 bites a day but in an area that is heavily infested with mosquitoes, people may receive dozens of bites per day. People who are infected with malaria may get bitten more frequently than uninfected people, either because they are sick and debilitated, and so are unable to brush off or kill mosquitoes, or because their fever or some odorants given off in their breath or from their skin makes them more attractive to mosquitoes. The probability that a bite will result in transmission of malaria from an infected person to a mosquito depends on the density of gametocytes in the blood but is probably small, in part because of the native immunity of the mosquitoes and in part because mosquitoes ingest only a small volume (several microliters) of blood per bite.

Now consider the factors that contribute to the number of people who become infected by one infectious mosquito. The latent period for parasite growth in mosquitoes—that is, the interval between the time a mosquito acquires gametocytes by biting an infected person and the time that she has sporozoites in her saliva and is herself infectious—is dependent on environmental temperature but is probably on the order of ten days to two weeks. Since the life expectancy of mosquitoes in the wild is only a couple of weeks (this, too, is quite variable, because it depends on temperature and other ecological factors, and differs for different mosquito species), individual mosquitoes are probably infectious for only a short time. The only infectious mosquitoes are those who acquired malaria parasites relatively early in life and who then survived to a relatively old mosquito age. Even in areas with high malaria transmission rates, only a small percentage of mosquitoes are infectious. Although mosquitoes produce a limited number of sporozoites, the probability of malaria transmission from infected mosquitoes to susceptible humans is high—sporozoites are injected directly into the blood, bypassing both the physical barrier and the immune defenses in the skin. All in all, the number of mosquitoes who become infected by biting one infected person is probably much greater than the number of people who become infected by a single infected mosquito. The number of mosquitoes that bite an infected person throughout the long course of an infection is orders of magnitude greater than the number of people who are bitten by an infectious mosquito during her brief infectious period.

9.5 Virulence of *P. falciparum*

As with other host–pathogen interactions, the virulence of *P. falciparum* depends on the tradeoffs between virulence and transmissibility, or R_0. There are many reasons why selection of *P. falciparum* for increasing R_0 has resulted in the evolution of virulent parasites. First, *P. falciparum* has evolved mechanisms to evade our immune defenses, and so people with *P. falciparum* infections remain burdened with parasites and infectious for long periods. Competition among strains in mixed infections may have led to within-host selection for the ability of *P. falciparum* to parasitize a large fraction of red blood cells and replicate to high densities in infected people. Moreover, all of the ways that *P. falciparum* causes infected people to be sick and debilitated—the spiking high fevers, the anemia, the consequences of cerebral malaria—enhance parasite transmission by making infectious people more attractive to mosquitoes and more susceptible to being bitten. The virulence of *P. falciparum*—the long periods of sickness and debilitation, and the deaths it causes—is a byproduct of selection for increased infectivity but the mortality caused by *P. falciparum* does not benefit the parasite. Deaths due to falciparum malaria are tragic but unselected consequences of the virulence of this pathogen.

The coevolution of malaria parasites with mosquitoes differs radically from their coevolution with humans. As mentioned earlier, the life span of mosquitoes is only slightly longer than the latency period required for sporozoite formation. Moreover, malaria transmission depends critically on the activity, and thus on the health, of infected mosquitoes. The theory of host–pathogen (or vector-pathogen) coevolution would predict that malaria parasites have undergone selection to have minimal effects on the life expectancy or activity of mosquitoes. In other words, malaria has evolved to be relatively benign in mosquitoes. This doesn't mean that the parasites have no effect on their mosquito vectors. If nothing else, the utilization of resources for parasite development must decrease the fitness of the mosquitoes. In addition, *P. falciparum* may increase the feeding activity of infected mosquitoes, which would increase parasite transmission but also increase the risk of mortality to the mosquitoes (Cohuet et al. 2010). Mosquitoes have evolved native immune defenses that increase their resistance to plasmodia and other parasites, which suggests that the evolutionary ancestors of these parasites decreased the fitness of the ancestors of mosquitoes.

Another way of looking at the relationships between humans, mosquitoes, and malaria parasites is that the parasites can appropriate much greater resources from humans than they can from mosquitoes. Even human infants weigh kilograms, while mosquitoes weigh only milligrams. Plasmodia have been selected to utilize the resources of their vertebrate hosts for their own replication and to use their arthropod vectors for sexual reproduction and for transmission from one host to another. An infected human might produce billions of gametocytes, while an infected mosquito probably produces only thousands of sporozoites. The different roles of humans and mosquitoes in the *P. falciparum* life cycle explain why we are referred to as hosts and mosquitoes as vectors of the parasite.

9.6 Evolution of *P. falciparum* and other malaria parasites

Plasmodia are in the phylum *Apicomplexa*, named for the apical complex, a specialized subcellular structure localized in the apical pole of the cells. The evolutionary ancestors of contemporary apicomplexans were presumably free living but at some point in their history, they became obligate intracellular parasites. Because the sexual phase of their life cycle is carried out in the intestinal tracts of arthropods, plasmodia are thought to be descended from gastrointestinal parasites in the evolutionary ancestors of insects. When blood-sucking insects arose, perhaps 150 million years ago, plasmodia evolved the ability to parasitize the vertebrates upon which these insects preyed.

Genomic and physiological studies are elucidating the evolutionary relationships between the many species of plasmodia. *P. falciparum* is closely related to *P. reichenowi*, a malaria parasite of chimpanzees and gorillas. *P. reichenowi* is more genetically diverse than *P. falciparum*. All *P. falciparum* strains that are now circulating in humans apparently originated from *P. reichenowi*, possibly from a single cross-species transfer. Whether this transfer came from chimpanzees or from gorillas is still a matter of controversy. The origin of *P. falciparum* has not yet been accurately dated (estimates range between 10 000 and 3 million years ago). Clearly, *P. falciparum* and *P. reichenowi* did not co-speciate with their primate hosts. Instead,

P. falciparum evolved from *P. reichenowi* long after the hominin and chimpanzee lineages diverged (Rich et al. 2009).

The species specificity of *P. reichenowi* and *P. falciparum* depends in part on the interaction of a parasite protein known as EBA-175 with glycophorins, sialic acid-containing glycoproteins on the surface of erythrocytes. The binding of EBA-175 to glycophorins is required for the entry of merozoites into red cells. Glycoproteins in other great ape species contain N-glycolyl-neuraminic acid (Neu5Gc) as their primary sialic acid. Some time after the hominin lineage arose, a deletion mutation inactivated the gene CMP-Neu5Ac hydroxylase (CMAH), which encodes the enzyme that catalyzes the synthesis of Neu5Gc from N-acetyl-neuraminic acid (Neu5Ac). Because we lack this enzyme, our glycoproteins contain Neu5Ac rather than Neu5Gc. EBA-175 from *P. reichenowi* binds to glycophorins containing Neu5Gc but not those containing Neu5Ac; as a result, *P. reichenowi* is not a human pathogen. In contrast, EBA-175 from *P. falciparum* binds to Neu5Ac-containing glycophorins. The inactivation of CMAH presumably spread in early hominin populations because the loss of Neu5Gc conferred resistance to *P. reichenowi* or an ancestral plasmodium species that also recognized Neu5Gc (Martin et al. 2005). Some time later, a mutation that enabled *P. reichenowi* to recognize Neu5Ac led to the evolution of *P. falciparum* as a human pathogen.

The different species of human malaria parasites did not evolve from a recent common ancestor that was adapted to humans. Instead, they evolved separately, in other species, and became adapted to humans by separate cross-species transfers. Although speciation of parasites within a single host species is possible, it probably does not happen often, since the gene flow that maintains host populations as a single species will usually be accompanied by gene flow between the parasite populations that infect them. Parasite speciation is most likely to occur when a host lineage speciates or when a parasite becomes adapted to a new host species, as in the case of *P. falciparum*. *P. malariae* and *P. vivax* are closely related to plasmodia that infect chimpanzees and other apes but are evolutionarily distant from *P. falciparum*. They most likely entered the human population from one of these other primates. Less is known about the origin of *P. ovale* but it also was presumably transferred into humans from another primate species. Finally, the newly identified human pathogen, *P. knowlesi*, appears to be entering the human population from macaque monkeys. There have evidently been a number of cross-species transfers of *P. knowlesi* from monkeys into humans but transmission of this pathogen from humans to mosquitoes has not yet been described.

9.7 Mosquitoes that transmit *P. falciparum*

The spread of *P. falciparum* requires a population of susceptible mosquitoes as well as a population of susceptible people. Dozens of mosquito species can transmit *P. falciparum*. These species differ in the density of mosquito populations, their breeding habits, their proximity to humans, and their anthropophilicity, their preference for feeding on humans rather than on other animals. Two mosquito species, *Anopheles gambiae* and *A. funestus*, are especially important vectors for transmitting *P. falciparum* in sub-Saharan Africa. *A. funestus* lives in the rain forest while *A. gambiae* breeds in standing bodies of water in cleared areas. Even though

P. falciparum has probably been a human (or hominin) parasite throughout much of human evolution, falciparum malaria as we now know it is thought to have arisen in the last 5000–10 000 years, as a consequence of the spread of agriculture through sub-Saharan Africa (Carter and Mendis 2002). When hunter-gatherer groups migrated through the rain forest, they could become infected with *P. falciparum* from *A. funestus*, but malaria was probably not a major disease for these groups. *A. funestus* populations are relatively small, they feed on other animals in addition to humans, and they have a short range and so don't migrate with migrating humans. The slash-and-burn agriculture that was used by Bantu peoples as they expanded through central Africa decimated the animal populations around human habitats that provided alternate blood sources for mosquitoes and created the clearings and standing bodies of water that led to the evolution of *A. gambiae*. *A gambiae* lives close to human population centers and is a highly anthropophilic species. *P. falciparum* is thought to have evolved into a more virulent pathogen in the ecological context in which it was being transmitted by *A. gambiae*.

P. falciparum is found only in areas where mosquitoes are abundant and are present throughout the year. In contrast, *P. malariae* and *P. vivax* are adapted to environments in which mosquitoes are present for only a few months a year. These parasites have evolved to be less virulent and to produce even more chronic infections than *P. falciparum*. When their sporozoites enter liver cells, some differentiate into yet another form, known as hypnozoites, which remain dormant in the liver for long periods and produce merozoites only for short periods each year. The production of hypnozoites is presumably an evolutionary adaptation that enables the parasites to survive during long mosquito-free intervals.

9.8 Effects of malaria on human evolution

Since malaria has been a major cause of infant and childhood mortality for thousands of years, it is not surprising that it has led to selection of alleles that confer resistance to the disease (Kwiatkowski 2005). Given that the morbidity and mortality from malaria result from growth of merozoites in red blood cells, it is also not surprising that many of the alleles that decrease the virulence of malaria infections affect red blood cell function. We have already discussed the thalassemia alleles that have spread in malarious areas, especially in the Mediterranean, because they increased resistance to *P. falciparum* malaria. Sickle cell hemoglobin (HbS), which we mentioned briefly in Chapter 1, is another well-known adaptation that increases resistance to malaria. HbS results from a mutation in the β-globin gene that leads to the substitution of valine for glutamic acid in position 6 of the β-chain of hemoglobin. People who are heterozygous for the HbS allele (that is, who have the AS genotype, where A denotes the gene for "normal" adult hemoglobin) are resistant to malaria. HbS probably promotes resistance because as plasmodia grow in red cells containing HbS, the cells sickle (i.e. assume a sickle shape) and are removed by macrophages in the spleen. Destruction of plasmodia by macrophages may also increase immune responses to the parasite. Evidently, the cost of increased red cell destruction is more than outweighed by the benefit of destruction of the parasites. People with the SS genotype, who are homozygous for HbS, suffer from sickle cell anemia. Their abnormally shaped erythrocytes have a short lifetime because they are taken up and destroyed in the spleen. In addition, their erythrocytes aggregate, plug up capillaries, and block blood flow to tissues. Historically most children with sickle cell anemia died in childhood. Despite recent medical advances, the SS genotype still results in greatly decreased evolutionary fitness.

The HbS allele has spread in regions where malaria is prevalent because the fitness benefits of malaria resistance outweighed the fitness loss of sickle cell anemia. The HbS allele is now widespread in sub-Saharan Africa and in some regions of Asia. Haplotype analysis, together with studies of the geographic distribution of HbS, suggests that the HbS mutation arose independently several different times. These mutations then spread from their sites of origin by natural selection and gene flow. The steady-state frequency at which the HbS allele will be maintained depends on the mortality rate from malaria. Given current estimates of malaria mortality in sub-Saharan Africa, the fitness of the AS genotype is perhaps 10–20% greater than that of the AA genotype (Hedrick 2004); this difference in fitness confers a large selective advantage for heterozygous people. As long as malaria remains prevalent and the AS genotype has increased fitness, the HbS allele will be maintained in the population and children will continue to suffer from sickle cell anemia. In the United States, where economic development has led to the eradication of malaria, the AS genotype no longer confers increased fitness and the frequency of the HbS allele is decreasing.

It is curious that the HbS allele has spread in sub-Saharan Africa and Asia, while the thalassemia alleles have spread in the Mediterranean. It was probably just a historical accident that these alleles arose where they did. But there appear to be negative epistatic interactions between HbS and thalassemia alleles. People who are heterozygous both for HbS and for thalassemia have decreased fitness (Penman et al. 2009). Because of this negative epistasis, once either of these alleles has become prevalent in a population, the other will not be able to spread into it.

Many other mutations that affect hemoglobin structure or other aspects of red blood cell function appear to have spread because they, too, increase resistance to malaria. Some of these mutant alleles, such as HbC, confer malaria resistance without causing any negative consequences, while others, such as mutations in glucose 6-phosphate dehydrogenase, may also cause disease. Several hundred thousand people a year may die from the diseases caused by mutant alleles that spread because they increased resistance to malaria. These deaths may be thought of as indirect effects of malaria.

Recall that glycophorins serve as receptors that enable *P. falciparum* to bind to and enter red blood cells. A different protein on the surface of red blood cells, known as the Duffy antigen, serves as a receptor for the malaria parasite *P. vivax*. West African populations and the African-American descendents of these populations have a high prevalence of the Duffy negative allele, which presumably spread because it confers resistance to *P. vivax* or a related parasite. As we shall see later, this allele may increase the risk of asthma or other allergic diseases.

Mutations in genes of the immune system are also important in increasing resistance to malaria (Kwiatkowski 2005). People in malaria-endemic areas of sub-Saharan Africa have a high frequency of several MHC or HLA alleles, genes that play key roles in the recognition of and response to pathogens. One is an allele known as B53. People with B53 appear to have only 50% as great a risk of developing severe malaria and dying as do people without this allele. B53 binds a peptide that is produced from a specific protein made by the liver stage of *P. falciparum*. Binding of B53 to this peptide on the surface of infected hepatocytes results in the killing of these cells by cytotoxic T cells. B53 is present in about 25% of the population in the Gambia while HbS while HbS is present in about 13%. So, even though it provides less protection than HbS (50% vs 90%), it is responsible for a significant share of the community resistance to *P. falciparum*. The distribution of B53 and its specific effect on resistance to malaria provides some of the best evidence for the idea that MHC polymorphisms are produced by the selection pressure of pathogens.

9.9 The future of malaria

For the reasons we discussed earlier, natural selection has caused *P. falciparum* to become a virulent pathogen. Unless ecological conditions in sub-Saharan Africa change dramatically, there is no reason to expect that the parasite will evolve to become less virulent. Aggressive campaigns to detect and treat patients with malaria have significantly reduced mortality from the disease. Several organizations have declared the goal of eliminating malaria. Whether this is a realistic goal remains to be seen. There are many innovative suggestions for the development of vaccines, the use of genetically modified mosquitoes, and other strategies to decrease the transmission of malaria. One would not want to bet against the creativity and ingenuity of scientists and the power of public health practices to prevent disease. On the other hand, one would not want to bet against the power of natural selection to overcome any human interventions. Clearly, evolutionary considerations should underlie any interventions designed to decrease or eliminate malaria. If at any time about 100 million people are infected with *P. falciparum* and if these people on average are home to about 100 million merozoites, there are on the order of 10^{16} *P. falciparum* merozoites in the world. This population must harbor an uncountable number of mutations. Designing an intervention that would eliminate all of these merozoites, with all of their mutations, seems a daunting task. The population of sporozoites is much smaller than the population of merozoites. Interventions that target this life cycle stage seem more likely to be successful. And even if we can eliminate *P. falciparum*, is *P. knowlesi* or some other plasmodium species likely to take its place? All of the emphasis on high-tech scientific interventions shouldn't cause us to lose sight of the fact that malaria is a disease of poverty (Carter and Mendis 2002). Economic development may be as difficult to achieve as an effective malaria vaccine but it is another approach to reducing or eliminating the burden of malaria. For the foreseeable future, unfortunately, malaria is likely to remain a major cause of disease and death.

10

Gene–culture coevolution: lactase persistence

10.1 Introduction

Humans are among the most successful species on Earth. The human population has grown to be incomparably larger than the populations of other primates (the chimpanzee population, for example, is thought to be about 300 000), we have colonized a wide range of habitats across almost the entire globe, and we have domesticated many other species and bred them to serve our purposes. Our unprecedented success is the result of our having evolved to create and occupy what has been called a "cultural niche," a niche in which learning from others is essential (Boyd et al. 2011).

Our capacity for culture did not arise full-blown from nothing. Like all of our other traits, this capacity has an evolutionary history. Many other species, including chimpanzees, learn by imitating the behavior of other members of their communities. Some time during hominin evolution, our ancestors evolved the abilities to modify and accumulate socially transmitted information, and to represent it symbolically. Our capacity for culture must have evolved because it enhanced the fitness of our evolutionary ancestors. The most attractive hypothesis to account for the evolution of social learning is that it enables organisms to function in changing environments (Richerson and Boyd 2005). The Pleistocene epoch (now dated from about 2.6 million to 10–12 000 years ago), during which much of hominin evolution took place, was the time of the most recent Ice Age, when there were dramatic oscillations in global temperatures. Changes in temperature must have caused changes in our ancestors' need for shelter, in their food supply, and in the predators with which they had to cope. The ability to acquire information by imitation and learning from others would have greatly enhanced our ancestors' ability to survive in such a changing environment. The stone tools and other implements that our Pleistocene ancestors made, as well as their use of fire, is evidence of their well-developed ability to learn and transmit social information (Wrangham 2009).

As we discussed in Chapter 2, our species, *Homo sapiens*, arose in East Africa about 150–200 000 years ago. These early *Homo sapiens* are sometimes referred to as anatomically modern humans. Their skeletons are similar to ours but, as judged by the tools they made, their behavior was similar to that of earlier species from which they evolved. Behaviorally modern humans, who made and used a much richer array of artifacts, seem not to have arisen until roughly 90 000 years ago (Tattersall 2009). By this time, our ancestors had developed what we

Evolution and Medicine. First Edition. Robert L. Perlman © Robert L. Perlman 2013.
Published 2013 by Oxford University Press.

would recognize as evidence of human culture; they made symbolic art and jewelry as well as tools. The apparent delay between the evolution of anatomically modern and behaviorally modern humans is puzzling. Our capacity for culture requires well-developed cognitive abilities. The transition from anatomic to behavioral modernity may have required genetic evolutionary changes in the human population that gave us these cognitive abilities (Klein 2009). But demographic factors play a major role in the development and maintenance of culture, as they do in genetic evolution (Powell et al. 2009). Beneficial cultural variants, like beneficial mutations, are rare and so are more likely to occur in large populations. And useful cultural practices, again like beneficial alleles, are less likely to be lost from large populations than from small. Behavioral modernity may have arisen when populations became large enough to support the accumulation and modification of cultural innovations. However our modern capacity for culture arose, it seems likely that our ancestors had developed this capacity by the time they left Africa and began colonizing other continents.

For most of human evolutionary history, our ancestors lived as hunter-gatherers, or foragers. Even after they had developed a capacity for artistic and technological innovation and for the cultural transmission and modification of these innovations, they had relatively little impact on their environment. They remained, in Jared Diamond's term, the "third chimpanzee" (Diamond 1992). We have seen how our ancestors' way of life changed dramatically some 12 000 to 10 000 years ago, at the end of the Pleistocene epoch, as foraging groups in several parts of the world began to domesticate plants and animals, develop agriculture, and live in permanent settlements. And we have discussed how these changes led to a rise in infectious diseases as well as to an increase in population growth.

The transition from a nomadic to a settled lifestyle had profound ecological and evolutionary consequences in addition to those we have already discussed. As our ancestors became agriculturalists and then urban dwellers, they began to alter their environment in an unprecedented manner. Since that time, the environments in which humans have developed, lived, and reproduced have increasingly become culturally constructed or man-made environments. In the words of John Odling-Smee and his collaborators, we have become unprecedented "niche constructors" (Odling-Smee et al. 2003). As cultural practices have changed and are continuing to change our environment, they have changed the selection pressures that acted on our more recent human ancestors and that are acting on us. Our culturally constructed environment has changed the fitness of different human genotypes and so has altered the course of human evolution. Conversely, the genetic makeup of human populations may influence the cultural information and practices they transmit and adopt. Human beings and human societies are thus the product of gene–culture coevolution (Boyd and Richerson 1985; Durham 1991). The coevolution of dairying and of lactose metabolism provides a good illustration of this coevolutionary process.

10.2 Milk consumption

American culture has long encouraged a belief in the nutritional benefits of milk for everyone, adults as well as children. Acting on this belief, Americans have been major consumers of milk. Although milk consumption in the United States has been declining since the Second World War, Americans on average still consume over 20 gallons of cow milk per year, not counting butter or cheese. Globally, per capita milk consumption ranges from almost 50 gallons per year in Finland to 29 in the United Kingdom and 2 in China. Ironically, as domestic milk consumption in the United States began to decline, milk became an important part of our foreign aid programs. During the 1950s and 1960s, the U.S. government together with private aid groups shipped millions of tons of powdered milk to underdeveloped countries. Soon, however, there were reports that milk was making people sick and that the recipients of our foreign aid were refusing to drink it. In 1971, the medical anthropologist Robert McCracken summarized these reports, many of which were accounts he received from Peace Corps volunteers. In Colombia and Guatemala, people did not drink the milk but used it to whitewash their homes, and among the Kanuri people in West Africa, the milk was rumored to contain evil spirits, presumably because it made them sick (McCracken 1971).

At about the same time, clinical studies demonstrated that many adults who complained of "lactose intolerance"—that drinking milk caused abdominal pain, diarrhea, and flatulence—were found to have what was then called "lactase deficiency," a marked decrease in lactase activity in their small intestine. The frequency of lactose intolerance and lactase deficiency showed striking differences among population groups, being much higher in African-Americans and Asians than in European Americans (Bayless and Rosensweig 1966; Cuatrecasas et al. 1965).

In the 1970s, McCracken and Frederick Simoons independently proposed the "cultural genetic" hypothesis, the hypothesis that high lactase activity and lactose tolerance were prevalent in societies that had a long history of dairying and milk use, whereas lactase deficiency and lactose intolerance were common in societies that did not have a history of dairying (McCracken 1971; Simoons 1978). Research since that time has confirmed and extended this hypothesis.

10.3 A brief history of animal domestication and dairying

Animal domestication began in Western Asia, probably somewhere in or near what is now Iraq. Domestication of sheep and goats began about 9000 BCE, followed soon after by domestication of pigs and cattle (Gerbault et al. 2011). The practice of domesticating animals spread rapidly into Anatolia (the Asian portion of present-day Turkey) and then more slowly into Europe. As we discussed earlier, animal domestication spread primarily by the migration of farming populations into territory that was previously used by foraging people. Agriculture supported greater rates of population growth than did foraging and so agriculturalists were able to outcompete foragers for land and other resources. Animal domestication may also have spread to a lesser extent by cultural diffusion, in which foragers adopted the practices of their farming neighbors. Domesticated animals were probably kept originally for their meat but milking may have begun soon after animals were domesticated. Early farmers may have begun milking cows because the total energy yield in a cow's milk is much greater than that in her meat. In other words, it is more efficient to convert the calories in grass into milk than into meat. In addition, feeding animal milk to infants might have increased fertility by enabling early weaning, thereby reducing the duration of lactation-induced anovulation. Analysis of organic residues recovered from pottery shards suggests that dairying and milk use were established in Anatolia by 6500 BCE (Evershed et al. 2008). Today, almost all peoples who keep cows use milk or milk products. Only a few societies have cattle but do not milk them.

Because of the importance of cows and cow milk in Europe and the United States, we know more about the history of cattle domestication and milking than we know about the domestication and milking of other species. Our domesticated cattle were derived from the now-extinct aurochs. Genetic evidence suggests that European cattle and the zebu or humped cattle found in South Asia were domesticated independently, from different subspecies of aurochs (Loftus et al. 1994). Aurochs may also have been domesticated in Northeast Africa. Despite our understandable interest in cow milk, we shouldn't forget that milk from many other mammalian species has been important historically as well as today. Almost a dozen different domesticated animals have been used for milk. For example, Arab Bedouins domesticated camels and have traditionally relied on camel milk.

10.4 The evolution of lactose synthesis and metabolism

Milk is the secretory product of mammary glands, the eponymous mammalian organs. Mammary glands evolved from dermal secretory glands in the egg-laying mammal-like reptiles that gave rise to mammals. The original function of mammary gland secretions is unclear. These secretions may have provided nutrition for offspring, helped to protect them against pathogens, or promoted communication and bonding between mothers and their offspring. Whatever this original function may have been, genomic studies showing the conservation of caseins and other milk protein genes suggest that lactation, the secretion of milk for the nourishment of newborns, was already developed by the time of the origin of mammals about 200 million years ago (Lefevre et al. 2010).

Milk is a complex fluid containing (among many other things) protein, fat, carbohydrate, and calcium. Lactose, the characteristic carbohydrate in milk, is a disaccharide made up of two monosaccharides, glucose and galactose, connected in a β-galactoside linkage. Although several plant species produce small amounts of lactose, mammalian milk is by far the most abundant source of lactose in nature. Storage of carbohydrate in milk as a disaccharide rather than a monosaccharide may have evolved because it decreased the osmotic load that accompanied storage. Sucrose is the most common disaccharide in nature. Mammals may have begun to make and store lactose rather than sucrose or another disaccharide simply because lactose synthesis arose first. Acquisition of the ability to synthesize lactose required only a small evolutionary change, one new enzymatic activity resulting from the evolution of one new gene. Mammary glands, like many other tissues, contain the enzyme galactosyl transferase, which catalyzes the transfer of galactose from uridine diphosphate galactose to a variety of acceptor molecules, especially glycoproteins and glycolipids. During late pregnancy and throughout the period of lactation, mammary glands synthesize the protein α-lactalbumin. α-Lactalbumin binds to galactosyl transferase and changes its substrate specificity. In the presence of α-lactalbumin, galactosyl transferase can utilize glucose as an acceptor for galactose and so can catalyze the synthesis of lactose. Together, galactosyl transferase and α-lactalbumin comprise the enzyme "lactose synthase."

α-Lactalbumin is similar in structure to lysozyme, an anti-bacterial enzyme that binds to and catalyzes the hydrolysis of carbohydrates in bacterial cell walls. Lysozyme is produced by many secretory tissues, almost certainly including the skin secretory glands that were the evolutionary precursors of mammary glands. The lysozyme gene underwent duplication at some time during the evolution of mammal-like reptiles. Subsequently, one copy of the duplicated gene accumulated mutations that changed the function of its protein product such that it now bound to galactosyl transferase and could participate in the synthesis of lactose; i.e. it became α-lactalbumin. The presence of the α-lactalbumin gene in all mammalian lineages is further evidence that lactose synthesis was already established by the time that mammals arose. Pinnipeds—Pacific seals, sea lions, and walruses—are said to be the only mammals whose milk does not contain lactose (Kretchmer 1972). The α-lactalbumin gene in pinnipeds has acquired other mutations that abolish lactose synthesis.

When lactose is ingested, it is first metabolized to glucose and galactose. Just as lactose synthesis essentially occurs only in mammalian mammary glands, lactose metabolism in nature is almost completely restricted to the mammalian gastrointestinal tract. Only epithelial cells in the intestine and some of the bacteria that normally inhabit the intestine have the ability to metabolize lactose. Evolution of the ability to convert lactose into glucose and galactose also required only one new enzymatic activity, lactase. Lactase is localized to the outer or apical surface of epithelial cells in the small intestine and is most highly expressed in the jejunum. When lactose is ingested, the glucose and galactose that are produced by lactase activity are transported across the intestine and then to the liver, where they enter the pathways of intermediary metabolism. Intestinal lactase has two active sites; one catalyzes the hydrolysis of lactose while the other catalyzes the hydrolysis of other carbohydrates. The enzyme is usually assayed by measuring the hydrolysis of the plant glucoside phlorizin—hence its technical name, lactase phlorizin hydrolase—but the natural substrates of phlorizin hydrolase are thought to be glucocerebrosides, complex glycolipids found in plant and animal cell membranes. Phlorizin hydrolase is widespread among vertebrates but lactase activity is found only in mammals. Lactase may have evolved by duplication of the active site sequence of a pre-existing phlorizin hydrolase gene, followed by mutations in one of these duplicated sequences that resulted in the ability to hydrolyze lactose (Mantei et al. 1988).

Ingested lactose that is not metabolized in the small intestine passes into the colon, where it is metabolized by intestinal bacteria. Of the hundreds of bacterial species in the intestinal microbiome, perhaps several dozen, including bifidobacteria, lactobacilli, and E. coli, have evolved the ability to metabolize lactose. β-Galactosidase, the bacterial enzyme required for lactose metabolism, catalyzes the same chemical reaction as does intestinal lactase but differs from lactase in structure, enzymatic properties, and regulation; the two enzymes evolved independently. Glucose and galactose are not transported across the colonic epithelium.

When these sugars are produced by bacterial β-galactosidase, they are metabolized further by intestinal bacteria and give rise to a variety of products, including three- and four-carbon acids, and hydrogen gas.

10.5 Lactase restriction and lactase persistence

Lactase itself is difficult to study, because it is a membrane-bound enzyme that is expressed only in intestinal epithelial cells. Lactase activity can be assayed directly in humans only by intestinal biopsy, an invasive procedure that for understandable reasons is rarely employed. The most commonly used clinical assay of lactase activity is measurement of hydrogen in expired air, which is a reflection of the lactose that was not metabolized by lactase and so was degraded by intestinal bacteria.

Most newborn mammals have high lactase activity during the nursing period, when lactose would normally be present in their diet, but lose it during or shortly after weaning and express low levels of lactase as adults, when they would not normally encounter lactose (Kretchmer 1972). This phenotype, in which lactase expression is restricted to the nursing period, is known as lactase restriction. The molecular basis for the loss of lactase after weaning is not yet well understood. As far as we know, lactase is the only gene whose expression is restricted to this period. Weaning is not accompanied by any dramatic neuro-chemical or hormonal changes that might lead to the loss of lactase expression. Mammalian lactase is not an inducible enzyme like the β-galactosidase in *E. coli*. Although nursing or feeding lactose may slow the loss of lactase activity, the decline in lactase expression is not due to the absence of dietary lactose and can't be prevented or reversed by feeding animals lactose. The patchy distribution of lactase on intestinal epithelial cells in adult animals is consistent with the hypothesis that the loss of lactase is due to the epigenetic regulation of lactase gene expression (Auricchio and Maiuri 1994). The gene is presumably turned off in stem cells in intestinal crypts by DNA methylation or some other epigenetic mechanism and the cellular phenotype of lactase restriction then spreads in a clonal fashion to the daughter cells.

The evolutionary explanation for lactase restriction is also not clear. It seems unlikely that the synthesis of a single enzyme in a single cell type would be sufficiently costly to decrease fitness and be selected against. A mutation that led to decreased lactase expression after weaning might possibly have spread by genetic drift some time early in mammalian evolution. One intriguing idea is that the loss of lactase activity in growing mammals is an adaptation that facilitates weaning and the transition from milk to non-milk food sources (Lieberman and Lieberman 1978). Newborn mammals rely exclusively on breast milk but at some point maternal milk is no longer sufficient for their optimal development. At that time, it becomes beneficial for the growing animals to seek alternate sources of nutrition. The symptoms that accompany the inability to digest lactose might be one of the stimuli that encourage offspring to abandon nursing and to seek out and try these other food sources.

The majority of humans have the phenotype of lactase restriction. They lose lactase activity some time during childhood and are not able to metabolize significant amounts of lactose as adults. The difficulty of measuring lactase activity has confounded attempts to determine the time course of the loss of enzyme activity in people with lactase restriction. It would be unethical to perform serial intestinal biopsies of children to measure lactase activity or mRNA levels. Only one set of studies has reported these measurements, made in children who were undergoing intestinal biopsy for other reasons. These were population studies, not serial measurements of individuals. Finnish children with lactase restriction exhibited a loss of lactase mRNA and a decline in lactase activity between roughly the ages of five and nine (Rasinpera et al. 2004; Rasinpera et al. 2005). A study of Sardinian children, in which lactase activity was estimated by measurement of breath hydrogen, reported similar results. About a third of children with lactase restriction showed decreased lactase activity (i.e. they had positive breath tests) by age three and all had lost activity by the time they were nine years old (Schirru et al. 2007). Earlier studies suggested that lactase activity might decrease at different ages in children in different populations. The time course for the decrease in lactase activity may well vary in response to genetic or environmental signals but as these studies were carried out under different conditions and with different methods, it is hard to draw firm conclusions about this.

Given that other mammals exhibit lactase restriction, this must have been the ancestral human phenotype. It is difficult to assess the "facilitation of weaning" hypothesis for lactase restriction in humans. As we discussed in Chapter 5, chimpanzees nurse their babies for 4–5 years whereas most contemporary forager populations wean their infants at 2½–3 years (Robson et al. 2006). Early weaning may have been an important aspect of human evolution, since it increased fertility rates and population growth rates. The time of weaning is apparently quite plastic, as it responds to ecological conditions such as the availability of other food sources for the weanlings. And as noted above, the time course for the loss of lactase activity may also be variable. Nonetheless, the available data suggest that infants in most human cultures would have been weaned long before they lost lactase activity, so that lactase restriction would not have played a significant role in weaning. Lactase restriction may be a vestigial trait that we inherited from an ancestral hominin species that weaned their offspring later than we do now.

In a few populations, notably those of Northern and Western Europe and their offshoot populations in Australia, New Zealand, South Africa, and the Western hemisphere, and also in some pastoral groups in Western Asia and North Africa, many people express lactase and retain the ability to metabolize lactose throughout life. These people are said to have the phenotype of lactase persistence. Because studies of lactose metabolism were initially carried out in European populations, lactase persistence was thought to be the "normal" phenotype, and adults who could not metabolize lactose were given the diagnosis of lactase deficiency. Lactase restriction is a more accurate description of this phenotype. (Primary lactase deficiency, in which people never express lactase, is a rare hereditary trait caused by mutations in the coding region of the lactase gene. It is different from lactase restriction.) Most human populations appear to be polymorphic for lactase restriction and lactase persistence (Itan et al. 2010). The frequency of the lactase persistence phenotype ranges from 10% or less in Southern China to greater than 90% in some Scandinavian populations.

Lactase persistence is caused by mutations in the enhancer region of the lactase gene. Several independent mutations have given rise to the lactase persistence phenotype in different populations. The primary mutation causing lactase persistence in Europeans is a C→T mutation 13.9 kb upstream from the lactase gene (C-13910T) (Enattah et al. 2002) but several other mutations in this region appear to be responsible for most of the lactase persistence in African populations (Tishkoff et al. 2007). The identified lactase persistence mutations do not account for the prevalence of the lactase persistence phenotype in many populations, and so other as yet unidentified mutations almost certainly exist (Itan et al. 2010). As in the case of malaria resistance, natural selection has resulted in convergent evolution, the spread of different mutations that result in a similar phenotype in different populations. These mutations maintain high levels of transcription of the lactase gene only on the chromosome bearing the mutation. Because we don't yet understand the molecular mechanism that underlies lactase restriction, we don't understand how these mutations result in lactase persistence. Presumably, lactase persistence is due to the loss of the putative epigenetic regulatory mechanism that would otherwise lead to a decline in lactase expression. Lactase persistence is inherited as a dominant trait. Evidently, lactase expression from one allele is sufficient to give heterozygous individuals the phenotype of lactase persistence. As we discussed in Chapter 3, beneficial dominant mutations would be expected to spread more rapidly than recessive mutations because they increase the fitness of everyone who carries them.

When adults with lactase restriction consume lactose, they may develop the clinical syndrome of lactose intolerance, which includes abdominal cramps, diarrhea, and flatulence. Recall that, in the absence of lactase, ingested lactose is largely metabolized by colonic bacteria. The products of bacterial lactose metabolism stimulate intestinal contraction and give rise to the symptoms of lactose intolerance. Because these symptoms depend on the amount and rate of lactose ingestion as well as on individual sensitivity to the products of bacterial lactose metabolism and probably also on the composition of the intestinal microbiome, they may not correspond closely to levels of intestinal lactase activity. Many people with lactase restriction can consume small amounts of lactose without difficulty. Conversely, other intestinal diseases can interfere with lactase activity and lactose absorption in people who have the lactase persistence genotype. For these reasons, there is an imperfect correlation between the lactase persistence/restriction genotypes and the lactose tolerant/intolerant phenotypes. Because lactose ingestion produces distressing symptoms in most people with lactase restriction, it would seem obvious that older children and adults with lactase restriction would consume significantly less milk than those with lactase persistence. Surprisingly, there are few firm data to support this idea (Bayless et al. 1975).

10.6 The coevolution of lactase persistence and dairying

The spread of lactase persistence in human populations is now recognized as an example—indeed, the prime example—of gene–culture coevolution (Durham 1991). Lactase persistence spread in some dairying cultures because people who retained the ability to consume and utilize animal milk after weaning had increased fitness in the environment shaped by the cultural practice of dairying and the availability of fresh milk. In terms of cultural niche construction, the spread of dairying modified the human environment and created a niche that increased the fitness of people with lactase persistence. The known lactase persistence mutations are embedded in extended haplotype blocks, which indicates that they have only recently increased in frequency as a result of natural selection. The dating of these mutations is consistent with the idea that they have spread within the last 5000–7000 years, since the time of animal domestication and dairying. Even though the alleles for lactase persistence have reached high frequencies in some populations (in some European groups, the frequency of the C-13910T allele is greater than 50%), they have not yet become fixed in any. The spread of these alleles reflects the pace of genetic evolution. As the lactase persistence alleles became more frequent, most copies of the ancestral lactase restriction allele would have been in heterozygotes, where they were shielded from natural selection because heterozygous individuals have the same fitness as homozygotes. Lactase persistence would have had to confer a fitness advantage on the order of perhaps 5–10% in order for lactase persistence alleles to have spread to their current levels since the advent of dairying (Bersaglieri et al. 2004). A selective advantage of this magnitude is among the strongest selective advantages reported in humans. Of course, the benefit of lactase persistence must have fluctuated throughout human history, depending on the need for people to rely on milk after weaning.

Although the benefits of drinking milk may strike us as obvious, it has been difficult to determine what exactly about milk drinking increased the evolutionary fitness of people who were able to do so. The early weaning of human infants is followed by a prolonged period in which children are dependent on their parents for sustenance. In societies that had domesticated cattle or other dairy animals, animal milk would have been a convenient and easily available post-weaning food. As discussed below, however, the general nutritional benefit of milk can be met by consuming cultured milk products that have reduced concentrations of lactose. Lactose itself may provide a specific nutritional benefit (Flatz and Rotthauwe 1973). In mice, lactose has a vitamin D-like action in promoting calcium absorption in the small intestine. This response to lactose is presumably an adaptation that increases calcium absorption in nursing pups. It isn't clear if this effect of lactose is significant in humans. If it is, then the ability to consume fresh milk, which contains both lactose and calcium, might have been especially important for the pastoral peoples who migrated to Northern Europe, where solar radiation and the endogenous production of vitamin D are low. On the other hand, lactose-induced calcium absorption would probably not have been important in pastoral African groups, who would have been exposed to high levels of UV radiation and would have been replete in vitamin D. Finally, apart from its nutritional benefits, fresh milk might well have been an important source of uncontaminated fluid, which might have been important in arid, desert environments. Thus, not only did lactase persistence arise independently in different populations, it may have enhanced fitness for different reasons in different environments. In developed countries today, lactase persistence is unlikely to confer a significant benefit, because most older children and adults have access to other foods and to clean water and do not need to rely on milk. In the past, however, and especially during times of nutritional shortages or when gastrointestinal diseases increased fluid loss and so increased fluid requirements, the ability to drink milk would have enhanced fitness. Lactase restriction may have been neutral in ancestral environments, or beneficial if it facilitated weaning, but it decreased fitness in a new environment where survival may have depended on the ability to consume fresh milk.

The geographical distribution of dairying is greater than is the distribution of lactase persistence (Holden and Mace 1997). All societies with a high frequency of lactase persistence are dairying societies but some cultures with a long history of dairying, such as those in Northern India and Pakistan, have a low frequency of lactase persistence alleles. Almost all dairying societies have developed the technology to process milk into cultured milk products, such as cheese, yogurt, kefir, or other yogurt-like beverages. These foods are produced by the bacterial metabolism or fermentation of milk. The bacteria that are used in cheese and yogurt production are normal intestinal bacteria (lactobacilli and other species) that express β-galactosidase and metabolize lactose. As a result of this bacterial fermentation, the lactose content of these foods is lower than in fresh milk. In cheese production, lactose is also removed with the whey. The lactose content of these cultured milk products is low enough that they can be consumed by most people with lactase restriction. Thus, there are two ways in which adults, and children after weaning, can derive the nutritional benefits of milk. If they have lactase persistence, they can drink fresh milk, and whatever their level of lactase expression, they can consume low-lactose cultured milk products. It is difficult to understand why some populations relied on fresh milk and so created an environment in which lactase persistence increased fitness, while other populations developed processes of milk fermentation. Perhaps in colder climates milk could be stored for longer periods and could be drunk while it was still fresh, while in warmer climates it quickly became rancid. Once the technology of milk fermentation spread in a society, however, lactase persistence would probably have conferred a smaller fitness advantage and so its increase in frequency would have been slowed. It is especially difficult to understand why some cultures in China and other East Asian countries domesticated cattle but apparently never developed the practice of milking. Many cultural traits must originate and spread (or fail to spread) for reasons that are independent of their effects on reproductive fitness.

In some cultures that have no history of dairying, such as the Yoruba in Nigeria and the Tadjiks in Afghanistan, up to 15–20% of adults retain the ability to metabolize lactose and so appear to have lactase persistence. The reason for the high frequencies of lactase persistence in societies where there has presumably been no selection for this phenotype is consistent with the idea that lactase restriction is a vestigial trait that is no longer beneficial to humans and that can easily be lost by genetic drift.

The concept of gene–culture coevolution entails not only that cultural practices affect the fitness of people with specific genotypes, but also that the genetic composition of a population influences the cultural practices that they adopt. Societies in which there is a high prevalence of lactase persistence are those societies in which dairying and milk drinking have spread, while societies with a high prevalence of lactase restriction have either not adopted the practice of dairying or have developed methods to produce low-lactose milk products. Although we don't have the historical data that would let us follow the coevolution of lactase persistence and of milk drinking over time, it is clear that genes and culture do coevolve.

Fortunately, lactase restriction is not a serious medical problem today, although it may still be an unrecognized cause of gastrointestinal symptoms in some people (Vesa et al. 2000). Nonetheless, it remains a source of continuing insights into our genetic and cultural evolutionary histories. And it has helped to clarify the confusion that can result from attempts to classify diseases as "genetic" or "environmental" (Hesslow 1984). In societies with a long history of dependence on fresh milk, most people have lactase persistence. In these societies, lactase restriction, or the inability to drink milk, may be thought of as a genetic disease. Indeed, this is how "lactase deficiency" was originally understood in the United States. In contrast, in cultures that do not utilize fresh milk and in which lactase restriction is prevalent, milk (or lactose) can be thought of as an environmental or culturally produced toxin—and in fact this is how some of the recipients of our foreign aid viewed the powdered milk we sent them. Lactose intolerance results from a culturally constructed environmental change that rendered old alleles detrimental. In this regard, it is similar to the man-made diseases we shall discuss in the next chapter.

11

Man-made diseases

11.1 Introduction

The idea that environmental change will change the fitness of genotypes is a foundational concept in evolutionary biology. We have discussed how changing agricultural practices led to the spread of *P. falciparum* malaria, which in turn increased the fitness of genotypes that were associated with resistance to malaria, and how the availability of fresh milk from domesticated animals increased the fitness of genotypes that were associated with lactase persistence. We stressed the idea that new alleles spread because they enhanced fitness but the converse is just as important. These environmental changes decreased the fitness of ancestral genotypes that were associated with susceptibility to malaria or with lactase restriction. The geneticist James Neel was among the first people to point out that, with respect to human health and disease, "Genes and combinations of genes which were at one time an asset may in the face of environmental change become a liability" (Neel 1962, p. 359).

The psychiatrist and Darwin biographer John Bowlby coined the term "environment of evolutionary adaptedness," or EEA, to refer to the environment to which our evolutionary ancestors became adapted through natural selection (Bowlby 1982, p. 50). Bowlby himself was interested primarily in human psychological traits that probably arose relatively recently, in the Pleistocene epoch, during which time the genus *Homo* and our species, *Homo sapiens*, arose. Many of the traits that characterize our species, including our increased cognitive abilities and our increased capacity for culture, evolved during this period. There was no single environment in which our ancestors lived or to which they became or were becoming adapted. Throughout our evolutionary history, our ancestors must have struggled to survive and reproduce in many different environments. Because environments were always changing and because of the constraints and tradeoffs we have discussed earlier, their adaptation was never perfect. The hypothetical EEA was not a Garden of Eden in which our ancestors lived without hardship or disease. Rather, there were many EEAs encompassing all of the environments our ancestors encountered throughout their evolutionary history and in which all of our myriad traits evolved.

We now live in an environment that differs in many important respects from the environments in which our evolutionary ancestors lived and to which they became more or less well adapted. We live in large, culturally diverse, and socioeconomically stratified communities of genetically unrelated individuals, we eat different foods, we are exposed to

Evolution and Medicine. First Edition. Robert L. Perlman © Robert L. Perlman 2013.
Published 2013 by Oxford University Press.

different sets of pathogens and toxins, and we have different patterns of physical activity than did our ancestors. Moreover, people in different human populations live in markedly diverse environments, in part because of geographical differences but also because their environments are shaped by different cultural traditions and practices.

In Chapter 2, we discussed the epidemiologic transition that is accompanying the demographic transition. As deaths from famine and infectious diseases decreased, and life expectancy increased, the burden of disease has shifted to chronic, noncommunicable diseases such as diabetes, coronary heart disease, and stroke, among many others (Armelagos et al. 2005; Omran 1971). The incidence of many of these diseases increases with age and their increasing prevalence is due in part to population aging. Some of these diseases involve loss of function and may be thought of as degenerative diseases. The main reason for the increasing prevalence of the major chronic diseases that now plague us, however, appears to be a mismatch between our present culturally constructed environment and our genetic inheritance, our heritage of alleles that enabled our ancestors to survive and reproduce under all of the various circumstances in which they found themselves. As we pointed out in Chapter 3, human populations are continuing to evolve in response to our changing environment. We are not trapped in Stone Age bodies, as some authors have suggested. Nonetheless, genetic evolution is a relatively slow process that is always lagging behind environmental—for us, largely man-made—change. Abdel Omran's term, man-made diseases, seems the most appropriate way to classify and think about these diseases (Omran 1971).

When we discussed genetic diseases, we pointed out that most cases of these diseases are caused by new or recent mutations. The mutant alleles that cause these diseases are kept at low frequency by mutation-selection balance, are lost by genetic drift, and are replaced by new mutations. In contrast, man-made diseases appear to arise from the interactions of old alleles with novel aspects of our environment. As Neel (1962) suggested, these are alleles that were an asset in ancestral environments but that have become a liability in the face of environmental changes. These may be ancestral or old alleles that are widespread in the human population, or less common alleles that arose after our ancestors migrated out of Africa and that influence susceptibility to diseases in specific populations. There has been great controversy over the roles of common and rare genetic variants in influencing susceptibility to these common man-made diseases. As the evolutionary geneticists Anna Di Rienzo and Richard Hudson have pointed out, "a single evolutionary model is unlikely to account for the genetic susceptibility to all common diseases" (Di Rienzo and Hudson 2005). We shall discuss several man-made diseases, or risk factors for disease, that are increasing in contemporary populations because of man-made, culturally constructed changes in our environment.

11.2 Diet, obesity, and diabetes

When he pointed out how environmental change may cause previously beneficial genes to become detrimental, Neel (1962) was developing a hypothesis to account for the increasing prevalence of adult-onset (insulin-resistant, or Type II) diabetes mellitus in Western societies.

His influential proposal, known as the "thrifty genotype" hypothesis, was based on the idea that our hunter-gatherer ancestors may have been exposed to recurrent periods of feast or famine. In an environment in which food intake was irregular, Neel argued, "a 'thrifty' genotype, in the sense of being exceptionally efficient in the intake and/or utilization of food," would have enhanced fitness (Neel 1962). Today, when food is plentiful, this genotype predisposes people to develop diabetes. Specific aspects of Neel's hypothesis remain controversial. It isn't clear if he was talking about common ancestral alleles that spread early in our evolutionary history or newer alleles that spread in selected populations, such as the Pima Native Americans who have experienced recent population declines or bottlenecks from famine. Anthropologists now believe that foraging populations had more stable food supplies and were at less risk of famine than were early agricultural societies. Genetic studies are identifying alleles that are associated with increased susceptibility to diabetes but the results of this genetic research are not yet well correlated with physiological studies showing how these alleles affect metabolism and actually contribute to a thrifty genotype. Despite these uncertainties, Neel's proposal that the prevalence of diabetes (and, by extension, other chronic, noncommunicable diseases) in our society is due to environmental or lifestyle change is now widely accepted.

The medical anthropologists S. Boyd Eaton and Melvin Konner also called attention to the importance of environmental change in increasing the burden of disease in contemporary societies (Eaton and Konner 1985). Eaton and Konner analyzed the diets of contemporary foraging populations and surveyed archeological evidence of early human diets such as tooth morphology, stone tools, and food remains found at sites of human occupation to infer what they called the Paleolithic Diet. (The Paleolithic era, or Old Stone Age, is defined in terms of the use of stone tools rather than in terms of geology. It corresponds roughly to the Pleistocene epoch.) Their retrodicted Paleolithic diet differed from current American diets in many respects: 1) protein was a greater percentage of caloric intake while fat was a smaller percentage; 2) the ratio of polyunsaturated to saturated fatty acids was higher; 3) intakes of fiber, calcium, potassium, and ascorbic acid were higher; and 4) sodium intake was lower. Our Paleolithic ancestors probably did not consume cereal grains at all. Moreover, because these ancestors engaged in more strenuous physical activity than we do, they had greater caloric requirements. Eaton and Konner pointed out that modern foragers are relatively free of obesity, diabetes, and heart disease, and proposed that the differences between our current diet and the diets of our Paleolithic ancestors contributed to the high incidence of these diseases in Western populations. Eaton and Konner's proposal is complementary to Neel's. Whereas Neel was concerned with alleles that were rendered deleterious by environmental change, Eaton and Konner focused on some of the specific environmental changes that may have rendered these alleles deleterious.

Just as there was no single EEA, there was no single Paleolithic diet. The foods that were available to our Stone Age ancestors must have varied greatly, depending on the local ecology, the season, and the changing climate that characterized the Pleistocene epoch. We evolved as omnivores and are able to eat a wide range of foods. The diets of contemporary foraging populations illustrate the great variation among human diets. For example, the Gwi people in Botswana subsist largely on plant foods and obtain less than 30% of their caloric intake from meat. In contrast, meat provides more than 90% of the caloric needs of Eskimo populations (Kaplan et al. 2000). There must have been comparable variations among ancestral human diets. Our modern diets evolved over time. Some changes, such as the increased consumption of grains, began with the agricultural revolution, while others, such as the increased use of refined sugar, are more recent (Cordain et al. 2005). If there had been strong selection for alleles that increased fitness in the presence of cereals or other aspects of the post-agricultural revolution diet, they too, like alleles for lactase persistence, should have spread in farming populations. Only the dietary changes that have occurred in the last several centuries are too new to have led to selection for genetic adaptations to them.

Evolutionary considerations aside, there are good physiological reasons for believing that some aspects of our modern diet and lifestyle do increase the risk of obesity and diabetes. Of course we gain weight because our caloric intake is greater than our caloric expenditure. But the development of obesity and diabetes is more interesting than this simple thermodynamic relationship may imply. Consumption of sucrose and of high-fructose corn syrup, in particular, has been linked to the rise in obesity and diabetes. Fructose increases lipid storage and may also increase appetite and reduce physical activity, and so may play an especially important role in the genesis of obesity (Wells and Siervo 2011). Because of their diets and their physical activity, people in foraging societies have more muscular and less adipose body compositions than do people in economically developed countries. Increased adiposity may also cause insulin resistance and may be one reason why obesity is a risk factor for diabetes (Eaton et al. 2009). Finally, diet can alter the composition of the intestinal microbiome. The microbiome of obese people differs from that in non-obese people. It is enriched in bacteria of the phylum Firmicutes, which are able to metabolize ingested carbohydrates and provide additional energy, which may also contribute to obesity (Kallus and Brandt 2012). Thus, in addition to providing excess calories, our modern diets may affect appetite, physical activity, body composition, and our microbiome. Even though we don't yet understand the relative importance of all of these factors or the interactions among them, it seems clear that recent changes in our diet play a central role in the increasing incidence of obesity and diabetes.

Investigating the role of diet in disease is difficult for many reasons. Evidence-based medicine was designed to evaluate the short-term effects of drugs or other medical interventions; it was not meant to assess the long-term effects of lifestyle changes or choices such as diet. Dietary studies can never have the credibility of the placebo-controlled, double blind studies that have become the "gold standard" of evidence-based medicine (Goldenberg 2009). Dietary studies also engender suspicion because diets often turn into fads. We may have to rely on a combination of epidemiological investigations, physiological studies, and common sense. Perhaps the best we can hope to do is follow sensible guidelines that are informed but not driven by evolutionary considerations, such as Michael Pollan's food rules: "Eat food. Not too much. Mostly plants" (Pollan 2008).

It is also difficult for individuals to implement dietary recommendations. Whereas taking pills is an individual activity, eating is communal. We eat what our friends and family are eating and, unless we are farmers, we eat what is commercially available and affordable. While we can educate and encourage people to eat healthier diets, such individual education and encouragement is unlikely to be successful unless it is supplemented by societal changes in food production, processing, and distribution (Lustig et al. 2012; Wells 2012).

11.3 Salt intake and hypertension

Hypertension, defined as a resting blood pressure usually greater than 140/90 mmHg in adults, is a risk factor for many diseases, including congestive heart failure, myocardial infarction, stroke, and kidney disease. Hypertension is thought to affect approximately 1 billion people worldwide. It is among the most important risk factors that contribute to the global burden of disease, exceeded only by undernutrition and unsafe sexual practices (Ezzati et al. 2002). There is now overwhelming evidence that salt intake is a major risk factor for the development of hypertension (He and MacGregor 2009). The level of salt in Western diets has increased dramatically in the last several thousand years. Hypertension is another clear example of a man-made pathological process, one resulting from a mismatch between our genetic heritage and our current environment.

Our relationship to salt has a long and interesting evolutionary history. Vertebrates evolved in a brackish or marine environment. Early vertebrates faced the physiological challenge of excreting the excess salt they took in from this environment in order to maintain physiological concentrations of sodium and chloride in their extracellular fluids. Sharks have evolved rectal glands, which concentrate and excrete salt. Teleost fish secrete salt from their gills and their kidneys. Osmoregulation in teleosts is controlled in part by the steroid hormones cortisol and deoxycorticosterone, acting on an ancestral mineralocorticoid receptor. When vertebrates migrated from saltwater into freshwater and then onto land, they moved from salt-rich to salt-poor environments. Survival in these new environments depended on the evolution of mechanisms to promote the acquisition and retention of salt. Aldosterone, the major mineralocorticoid hormone in mammals, apparently appeared first in lungfish, the evolutionary ancestors of tetrapods. Aldosterone promotes sodium reabsorption in the kidneys and so functions to conserve sodium. The evolution of aldosterone synthesis seems to parallel the transition in the primary task of the mineralocorticoid osmoregulatory system from salt excretion to salt retention.

Primates and other terrestrial mammals evolved in environments in which conservation of salt remained essential for survival and so have retained aldosterone and other mechanisms for the renal conservation of sodium. In addition, mammals have evolved taste receptors that are sensitive to salty foods and that make these foods appetizing. When our hominin ancestors migrated out of the protected forested environment onto hot open savannahs, they had to adapt to this new environment. Among other things, they evolved a greatly increased capacity for sweating. But while sweating promotes heat loss, it also increases fluid and salt loss. And although aldosterone promotes sodium reabsorption in sweat glands as well as in the kidney, we still lose sodium in sweat. Increased sweating heightened the importance of renal sodium retention (Young 2007). As a result of our evolutionary history, our kidneys are poised to conserve salt (Adrogue and Madias 2007). With respect to salt, we do have a thrifty genotype.

For most of human history, the only salt our ancestors consumed was the salt naturally present in their food. (Because almost all of the sodium in our diets is in the form of sodium chloride, we can consider sodium intake and salt intake interchangeably.) Meat contains much more sodium than does vegetable food, and so the sodium intake of our Paleolithic ancestors would have depended heavily on their diets. It is likely, though, that their sodium intake was less than 1 gram or about 50 mmol/day (Konner and Eaton 2010). Some contemporary foraging populations still have average sodium intakes considerably less than 1 gram/day. There is no evidence for salt mining or gathering before about 6000 BCE and even this evidence is only anecdotal (Kurlansky 2002). Initially, salt was probably used primarily as a preservative rather than as a food additive. It is difficult to know how salt intake increased since the advent of salt mining. Salt intake in the eighteenth century is said to be as great as it is today but this claim is hard to reconcile with the observation that most of the salt in contemporary diets is added during processing of commercial foods. Average sodium intake in the United States is currently about 3.4 g or almost 150 mmol/day, more than three times what our ancestors most likely consumed.

Both among-population and within-population studies document an association between sodium intake and blood pressure. Hypertension is virtually absent in those few contemporary populations whose average sodium intake is less than 50 mmol/day (Intersalt Cooperative Research Group 1988). People in these populations show little if any age-related increase in blood pressure. On the other hand, populations with higher levels of sodium intake have an increasing prevalence of hypertension and an increasing rate of rise in blood pressure with age. Hypertension is prevalent only in populations whose average sodium intake is above 100 mmol/day; above this level, the prevalence of hypertension increases with increasing sodium intake. We don't have good information about the effects of sodium intakes between 50 and 100 mmol/day because there are few populations with intakes in this range. These differences in blood pressure are not due to genetic differences between populations, as people who migrate from low-salt societies to societies with high salt intake acquire the risk of hypertension of their new countries.

Correlations between sodium intake and blood pressure do not establish causation. Both population-based intervention studies and patient-based treatment trials, however, confirm the association between salt intake and blood pressure and establish that a sodium intake above 50–100 mmol/day is a risk factor for hypertension. Even in short term studies (30 days), reduced sodium intake lowers blood pressure whereas high-salt diets increase it (Sacks et al. 2001).

Many people who consume high amounts of salt do not develop hypertension. A high salt diet is necessary but not sufficient for the development of hypertension. As we should expect, there is genetic variation in our response to salt. A number of alleles have been identified that increase salt sensitivity, or the rise in blood pressure with increasing salt intake (Young 2007). Some of these alleles are ancestral alleles, which we share with chimpanzees and probably other primates. These alleles were almost certainly maintained in these other species and in our own ancestors because they promoted salt conservation in low-salt ancestral mammalian or primate environments. Other alleles that increase salt sensitivity are common in the human population but are not shared with chimpanzees. These alleles probably spread because they increased the fitness of early hominins in the face of the extra stress of the salt and water losses they experienced. The frequency of these ancestral and common alleles that increase the risk of hypertension confirm the hypothesis that we are adapted to a low salt environment and that the prevalence of hypertension is due to recent increases in sodium consumption.

Some alleles that influence salt sensitivity show important differences among human populations. These alleles show a latitudinal cline, being present at high frequency in equatorial Africa and declining in frequency at northern latitudes. These alleles were presumably common in ancestral African populations and decreased in frequency as populations migrated away from the equator where temperatures were lower, sweating decreased, and selection for salt conservation was relaxed. The prevalence of hypertension is much greater among African-Americans than among Americans of European descent. One of the many factors that contribute to this difference is an increased frequency of salt-sensitive alleles in African-Americans (Young 2007).

Although most studies of mineral intake and hypertension have focused on sodium intake, other ions, especially potassium, are also involved in blood pressure regulation and in the pathogenesis of hypertension (Adrogue and Madias 2007). In addition to being lower in sodium, ancestral diets were higher in potassium than are contemporary diets. Plant foods, and especially vegetables, have higher potassium contents than do cereals, meat, or dairy products, and so the potassium content of ancestral diets, like the sodium content, would have depended on the relative importance of animal and plant foods in these diets. Given the range of foods in ancestral diets, the molar ratio of dietary potassium to sodium in these diets was probably between 3 and 10, while today it is less than 0.4. Population studies have shown an inverse relationship between potassium intake and the prevalence of hypertension, and clinical trials have shown that potassium supplementation lowers blood pressure. The relative importance of the absolute amount of sodium ingested and the sodium/potassium ratio is a subject of ongoing research. Moreover, while sodium is consumed primarily as sodium chloride, potassium is consumed with organic anions that are metabolized to bicarbonate. As a result, ancestral potassium-rich diets gave an alkaline metabolic load, while contemporary sodium-rich diets promote metabolic acidosis. The metabolic acidosis resulting from chloride intake may also contribute to the pathophysiological effects of a high salt diet (Frassetto et al. 2001).

Blood pressure regulation is complex and the way salt loading increases blood pressure is also complex. Excess sodium intake, or an elevated sodium/potassium ratio, raises blood pressure, at least in part, because the failure to excrete all of our ingested salt and water results in an increase in blood volume. Sodium intake activates the sympathetic nervous system, which promotes vasoconstriction, and stimulates the secretion of a ouabain-like hormone from the adrenal cortex, which increases renal sodium reabsorption (Blaustein et al. 2012). Blood pressure increases with age, so that hypertension is a pathology of aging, and the diseases for which it is a risk factor increase in frequency with aging. There are many reasons why blood pressure might increase with age. Arteries and arterioles become stiffer and less compliant because of the cross-linking of elastin and collagen fibers in their walls. Renal function declines with age, leading to more salt and water retention. Again, however, the age-dependent rise in blood pressure that is almost ubiquitous in the United States and other economically developed countries is not seen in populations whose average salt intake is less than 1 gram or 50 mmol/day. Hypertension is a man-made pathology.

Just as our kidneys are poised to conserve salt, our cardiovascular, endocrine, and autonomic nervous systems are poised to maintain our blood pressure in the face of stresses that would otherwise cause hypotension. Hypotension, due to dehydration, blood loss, or sepsis, was probably a much more important cause of death for our evolutionary ancestors than was hypertension. Given their likely salt intake, our ancestors must have been virtually immune to hypertension. As a result, we have evolved more powerful mechanisms to maintain blood pressure in the face of stresses that would otherwise cause hypotension than to lower blood pressure in response to challenges that increase it.

Despite the gaps in our information, the evidence that salt intake contributes to hypertension, which in turn contributes to heart disease, kidney disease, and stroke, is now compelling. Denial that salt is harmful comes primarily from people in the food industry who have a financial incentive to promote salt usage. Official nutritional guidelines recognize the relationship between sodium intake and blood pressure. The U.S. Institute of Medicine recommends a daily dietary intake of 1500 mg of sodium, or 65 mmol/day, as an adequate intake for young adults. This level of sodium intake would provide ample sodium even for people who had extreme salt loss due to exercise and sweating, and carries little or no risk of hypertension. The Institute of Medicine further defines a "tolerable upper limit" of sodium intake as 2300 mg or 100 mmol/day, which is the threshold above which the risk of hypertension clearly increases. Unfortunately, the bulk of our sodium intake, perhaps 75%, comes from salt added to food during its processing and manufacture. Only small amounts are naturally present in food or added during cooking or at the table. Given our reliance on prepared and restaurant food, it is difficult for people in the United States to consume less than this tolerable upper limit. As with the dietary changes we discussed earlier, societal changes in food processing will be required to decrease salt intake. Several countries, including Finland, Japan, and the United Kingdom, have begun to reduce the salt content in prepared foods. There is every reason to believe that this reduction in salt consumption will lead to a decrease in the prevalence of hypertension but these interventions are too recent to test this prediction.

11.4 Elimination of old pathogens: the hygiene hypothesis

The prevalence of many allergic and autoimmune diseases, including asthma, multiple sclerosis, and juvenile (insulin-dependent, or Type I) diabetes mellitus, has been increasing since the mid-nineteenth or early twentieth century. Taken as a group, autoimmune diseases are now the tenth leading cause of death among women under age 65 in the United States (Walsh and Rau 2000). Of these, multiple sclerosis, rheumatic fever and heart disease, and systemic lupus erythematosus are the most important causes of death. As with some of the other trends we have discussed, these diseases began to increase first in economically developed countries but are now becoming more prevalent in developing countries. There is genetic variance in the susceptibility to these diseases. As with the alleles that increase the risk of hypertension, many of the alleles that increase susceptibility to allergic and autoimmune diseases are common, old alleles (Barnes et al. 2005). This observation, together with the rapid recent increase in the prevalence of these diseases and the relatively low concordance among monozygotic twins, indicates that these diseases are increasing because of recent environmental changes. These diseases, too, appear to be man-made diseases.

In 1989, the epidemiologist David Strachan reported that the prevalence of hay fever was inversely correlated with household size and was greater in older than in younger siblings. Strachan proposed what has come to be known as the "hygiene hypothesis," the hypothesis that the increasing incidence of hay fever—and, by extension, other allergic and autoimmune diseases—is due to improved hygiene and reduction in exposure to pathogens (Strachan 1989). Since that time, this concept has become supported by a wealth of epidemiological, experimental, and clinical evidence. Reduced exposure to pathogens is now widely accepted as playing an important role in increasing susceptibility to these diseases (Rook 2012).

Parasitic worms, or helminths, have played an important role in the development of the hygiene hypothesis. Helminths have parasitized and coevolved with mammals for much of mammalian history. Two phyla of helminths are important human parasites, nematodes (roundworms) and platyhelminths (flatworms). Until the twentieth century, most if not all humans were infected by worms. Even today, billions of people, mostly in tropical, economically developing countries, still suffer from these infections. The most prevalent helminths are the nematodes *Ascaris lumbricoides*, *Trichuris trichiura* (whipworm), and hookworms. Over a billion people are infected with each of these parasites. We evolved in, and many of us still live in, a wormy world (Stoll 1999).

Helminths were able to survive and reproduce in the small, mobile populations of our ancestors because they modulate the immune response against them in ways that enable them to cause long-lasting or recurring infections. The worms appear to produce regulatory molecules that act directly on our immune systems and they may also act indirectly by modifying the composition or activity of our intestinal microbiome. But we have coevolved with these parasites. Helminths were such a ubiquitous and consistent part of our ancestral environment that our ancestors evolved traits that optimized their fitness in their wormy worlds. Presumably, they evolved to down-regulate their immune responses when stimulated by worms because more active responses were damaging. Because of the way we coevolved with helminths, the normal development of our immune systems now depends on infection by these organisms. In today's hygienic environments, we are no longer exposed to or infected by these parasites and so our immune systems are more reactive than they would otherwise have been.

Some of the most compelling evidence for the hygiene hypothesis and specifically for the role of helminths in autoimmune diseases comes from studies of patients with multiple sclerosis (Correale and Farez 2011). Although there are some genetic loci that increase susceptibility to multiple sclerosis, the low concordance between sets of monozygotic twins and the observation that the children of immigrants acquire the disease risk of their host country suggest that environmental factors play a large role in the pathogenesis of the disease. Multiple sclerosis is thought to be triggered by a viral pathogen that elicits an immune response that then damages the nervous system but no known pathogens have increased in frequency or have the geographic distribution of multiple sclerosis. Early studies demonstrated that the prevalence of multiple sclerosis was correlated with levels of hygiene and socioeconomic development. More specifically, the prevalence of multiple sclerosis was negatively correlated with the prevalence of *T. trichiura*.

This epidemiologic evidence has now been supplemented with more direct clinical observations. Patients with multiple sclerosis who happened to be infected with any of a number of helminths showed a more benign course, with fewer relapses and slower progression of disability, than did uninfected patients. The more favorable clinical course of these patients was correlated with increased levels of immunosuppressive T cells and cytokines. Some of these patients required treatment for their parasitic infections. Treatments that decreased or eliminated the parasites led to exacerbations or flare-ups of their multiple sclerosis.

Finally, recent preliminary clinical trials showed that short term treatment of multiple sclerosis patients with *Trichuris suis*, a pig whipworm that is not a human pathogen, led to a reduction in the severity of their disease, as judged by magnetic resonance imaging and immunologic studies. These benefits were temporary and were not sustained after the treatments were discontinued (Fleming et al. 2011). Helminths are known to secrete a variety of molecules that down-regulate the immune system and that prevent autoimmune diseases in experimental animals (Cooke 2012). Whether *T. suis* itself or any of the immunosuppressive substances released by the parasite can be developed into a useful therapy remains to be seen. But these studies provide strong support for the hygiene hypothesis and important insights into the pathogenesis of multiple sclerosis. If this research does lead to effective therapy for patients with this debilitating disease, it will be one of the most important clinical benefits to have come out of evolutionary medicine.

Parasitic worms also seem to play a role in inflammatory bowel disease (Crohn's disease and ulcerative colitis) (Weinstock and Elliott 2009). There are historic and geographic correlations between improved hygiene and increased incidence of inflammatory bowel disease. The children of immigrants acquire the risk of inflammatory bowel disease associated with their host countries. And preliminary clinical trials using *T. suis* suggest a benefit to patients with inflammatory bowel disease. Again, while these preliminary results provide additional support for the hygiene hypothesis, whether they will lead to clinically useful therapies remains to be seen.

Asthma is another disease whose prevalence appears to be increasing as a result of improved hygiene (Barnes 2011). It is a cruel irony that the prevalence of asthma is especially high among children in poor urban neighborhoods. Presumably these children have increased susceptibility to allergic diseases because they have grown up in helminth-free environments but they live under conditions in which they have greater exposure to dust mites, cockroaches, and other agents that evoke allergic responses.

Improved hygiene does not by itself cause allergic and autoimmune diseases. There are genetic polymorphisms that affect the risk of developing these diseases. Improved hygiene creates an environment in which some old alleles increase susceptibility to these diseases. These genetic polymorphisms may have arisen and spread for a variety of reasons. A polymorphism in the gene that encodes the Duffy antigen is an instructive example. Recall that the Duffy antigen is a protein on red blood cells that serves as a receptor for the malaria parasite *P. vivax*. West African populations and the African-American descendents of these populations have a high prevalence of the Duffy negative allele, which presumably spread because it conferred resistance to *P. vivax* or a related parasite. The Duffy antigen has several functions; i.e. the Duffy gene is pleiotropic. In addition to serving as a receptor for *P. vivax*, the Duffy antigen binds and inactivates chemokines, proteins that play an important part in inflammatory and allergic reactions. People who carry the Duffy negative allele appear to be at increased risk of asthma because they lack this ability to bind and remove chemokines from the circulation (Barnes 2011). The prevalence of the Duffy negative allele contributes to the burden of asthma in African-American populations. In this case, and presumably in the case of other polymorphisms that affect the risk of acquiring allergic or autoimmune diseases, improved hygiene has rendered a previously advantageous allele deleterious.

As with other man-made diseases, there is still much that we don't understand about the diseases whose increasing prominence has resulted from improved hygiene. Despite the attention given to the role of helminths in shaping our immune systems, we don't understand when in development exposure to helminths is critical. Lack of helminth infections probably does not account for our increased susceptibility to all of our various allergic and autoimmune diseases. Reduced exposure to other pathogens and environmental factors such as vitamin D deficiency may also contribute to the prevalence of these diseases.

No one would want to return to the unhygienic and helminth-infested environments that characterized much of human history. As devastating as allergic and autoimmune diseases are to patients, their families, and the society, the burden of these diseases is outweighed by all the benefits of improved hygiene. On the other hand, our cultural obsession with avoidance of "germs," which is presumably an outgrowth of our evolved disgust response, may now have gone so far that it has become injurious. In addition to helping us understand the pathogenesis of these diseases, a deeper understanding of the role of microorganisms in shaping our immune responses may help us figure out how we can keep the benefits of improved hygiene while minimizing its deleterious side effects.

11.5 Hierarchical societies and socioeconomic disparities in health

Socioeconomic disparities in health are among the most troublesome and refractory problems in medicine. Simply put, poor, disadvantaged, or marginalized people have poorer health and shorter life expectancies than do rich or more privileged people. The cause of these health disparities is hotly debated. People who have poor health may be unable to work, or unable to work at higher skilled or higher paying jobs, and so poor health may sometimes lead to poverty, but this factor must account for only a small fraction of the association between poverty and health. In countries such as the United States, disparities in access to health care may contribute to health disparities but this too must be only a small part of the problem, because health disparities exist in countries such as the United Kingdom, which has a national health service and in which there is less inequality in access to health care. Absolute levels of poverty compromise health and longevity in the poorest countries in the world, where life expectancy is correlated with the per capita gross national product, but poverty alone does not account for the health disparities in developed countries. The epidemiologists Michael Marmot and Richard Wilkinson have been the most forceful proponents of the view that socioeconomic inequalities themselves are the cause of inequalities in health (Marmot 2004; Wilkinson 2001). Although their conclusions are controversial, they have assembled an impressive amount of data in support of their argument.

Some of the most comprehensive and best-documented studies of health disparities have been carried out in the United Kingdom. The U.K. Office for National Statistics classifies occupations into seven socioeconomic classes, ranging from "1. Higher professional and higher managerial" (senior government officials, physicians, scientists, etc.) to "7. Routine" (occupations such as bus drivers and domestic workers), and analyzes health disparities in terms of this occupation-based scale (Office for National Statistics 2011). Table 11.1 presents some data from an ongoing longitudinal study of 1% of the U.K. population. People in occupational class 1 have longer life expectancies than do those in class 7. In the period from 1982–86, men in

Table 11.1 Life expectancy at birth by socioeconomic class, United Kingdom

Occupational Class	Life Expectancy (years)			
	Males		Females	
	1982–86	2002–06	1982–86	2002–06
1. Higher management & professional	75.6	80.4	80.9	83.9
2. Lower management & professional	74.3	79.6	79.7	83.4
3. Intermediate	73.3	78.5	79.6	82.7
4. Small employers	73.6	77.8	79.1	82.6
5. Lower supervisory & technical	72.3	76.8	78.5	80.4
6. Semi-routine	71.3	75.1	78.1	80.6
7. Routine	70.7	74.6	77.1	79.7
Difference between Class 1 and Class 7	4.9	5.8	3.8	4.2

Source: Office for National Statistics 2011

class 1 had a life expectancy of 75.6 years while men in class 7 had a life expectancy of 70.7 years. In the period from 2002–06, life expectancy of these groups increased to 80.4 and 74.6 years, respectively. Despite the dramatic increase in life expectancies during this time, the gap in life expectancy increased from 4.9 to 5.8 years. Similar data hold for women, except that women have higher life expectancies than men and the socioeconomic gap is a little smaller for women (3.8 years in 1982–86, 4.2 years in 2002–06).

Health disparities are not simply restricted to the most disadvantaged members of a society. There isn't a cutoff in socioeconomic status such that people below some level have poor health and people above it have good health. Instead, there is a social gradient in mortality rates that runs through all socioeconomic classes. This gradient is not just a gradient in mortality from one specific cause but results from socioeconomic differences in mortality rates from a host of different diseases. In addition to the socioeconomic gradient in mortality rates, there is also a gradient in morbidity. People in lower socioeconomic classes exhibit earlier onsets of chronic diseases and disabilities (Chandola et al. 2007). In terms of morbidity as well as mortality, people in lower socioeconomic classes appear to age more rapidly than do people in higher classes.

Studies of health disparities in the United States complement those in the United Kingdom. Most research on health disparities in the United States has related health status to ethnicity or racial identity, since this information is routinely collected on census forms and other official documents. Race in the United States (used in the sense of socially constructed and self-defined categories) is a cruder measure of socioeconomic class than is the occupational measure used in the United Kingdom but it captures similar features of disparities in wealth, status, and power. The life expectancy of non-Hispanic white Americans is about five years greater than that of non-Hispanic blacks (currently about 79 vs 74 years). This disparity is not simply due to a difference in infant mortality rates, since life expectancy at age 50 is still two to three years higher for whites than for blacks. Age-specific mortality rates rise in parallel for whites and blacks but the curve for blacks is shifted to about seven or eight years earlier than for whites. For example, the age-specific mortality rate of 50-year-old blacks is about the same as that of 58-year-old whites. Similar to the socioeconomic class differences in health in the United Kingdom, black Americans show an earlier onset of chronic diseases and disabilities and appear to age earlier than do whites.

The U.S. National Center for Health Statistics has recently published an extensive analysis of socioeconomic status and health (National Center for Health Statistics 2012). In these analyses, socioeconomic status was assessed either by family income or by educational attainment. Many measures of health were correlated with these indices of socioeconomic status. Table 11.2 gives one example, life expectancy at age 25 as a function of education level.

Again, it is not just that the most poorly educated people have the lowest life expectancy. Rather, there are gradients in life expectancy as a function of education level. And these gradients persist, even as overall health improves. Over the period from 1996 to 2006, life expectancy for college graduates increased, while life expectancy for people without a high school diploma stayed the same or decreased. Taken together, data from these two countries (and from a host of others) provide compelling support for the hypothesis that poor, disadvantaged,

Table 11.2 Life expectancy at age 25, by sex and education level, United States

Educational Level	Life Expectancy at Age 25 (years)			
	Men		Women	
	1996	2006	1996	2006
No high school diploma	47	47	53	52
High school graduate	50	51	57	57
Some college	51	52	58	58
College graduate	54	56	59	60

Source: National Center for Health Statistics 2011

low status people age more rapidly than do rich, advantaged, higher status people and that there is a socioeconomic gradient in the rates of aging (Crimmins et al. 2009).

Recall that we have evolved mechanisms that alter our life history strategies in response to our assessment of the security or insecurity of our environment and of our anticipated mortality rate (Chapter 5). Signals which suggest that our environment is dangerous and that our lives may be "solitary, poor, nasty, brutish, and short" lead to the diversion of resources away from bodily maintenance into early reproduction and to coping with and surviving acute stresses. Hobbes's description of the human condition may not apply to foraging populations, as he suggested, but it does fit the lives of poor and disadvantaged people in modern societies. In Chapter 5, we discussed the developmental conditions that increase the risk of adult disease and early mortality. These conditions, which include poor nutrition and psychosocial stresses such as social isolation, lack of material resources, and fear of violence, as well as prematurity or low birth weight and exposure to toxins, are more prevalent in lower than in higher socioeconomic classes (Mackenbach and Howden-Chapman 2003). In the economic terms we discussed earlier, poor people are likely to be born with lower amounts of physiological capital and to expend this capital at a higher rate. Although socioeconomic gradients in rates of aging are distressing, they should not be surprising (Perlman 2008).

The physiological mechanisms that underlie our responses to environmental signals are still being investigated but they appear to involve neuroendocrine regulatory pathways, including the hypothalamic-pituitary-adrenal axis and the autonomic nervous system. Hormones such as cortisol and catecholamines suppress the immune system and reduce tissue repair. These hormones enhance our responses to stress at the cost of reducing somatic maintenance (Phillips 2007). Socioeconomic gradients in the levels of these hormones or in the reactivity of these neuroendocrine systems may contribute to gradients in life history trajectories.

Socioeconomic gradients in health behaviors are important proximate causes of health disparities. People of lower socioeconomic status are more likely to smoke and to engage in other risky behaviors, and less likely to take advantage of preventive health services, than are people in higher socioeconomic groups. Unfortunately, interventions designed to reduce disparities in health behaviors, such as better education and greater provision of health services, have been disappointing. We appear to have evolved psychological biases as well as physiological mechanisms that lead us to discount bodily maintenance and future health when we assess our environment as dangerous or stressful (Hill 1993). Improved education or increased availability of preventive health services, while all well and good, are not likely to alleviate the root causes of health disparities, which are disparities in wealth, status, and power, and the ways we have evolved to respond both physiologically and psychologically to the place we find ourselves in hierarchical, socioeconomically stratified societies.

11.6 Reducing the burden of man-made diseases

The prevention and treatment of man-made diseases brings us back to the differences between an evolutionary or population-based view of biology and a traditional medical or individual-based perspective that we discussed in Chapter 1. Current biomedical research is focused on identifying the genetic and phenotypic factors that increase an individual's risk of developing disease and on designing preventive therapies to decrease this risk. Of course this is a most important research strategy. Identification of risk factors can give insights into both the pathogenesis and the prevention of disease. The value of interventions that modify risk factors can be documented by controlled clinical trials. Reducing blood pressure does decrease the incidence of myocardial infarction and stroke, and does save lives. Clinical studies identify the people who are most likely to benefit from medical treatment and to whom treatment should be offered. Even though this treatment is preventive rather than curative or palliative, it conforms to the historic role of physicians in caring for their individual patients. This caring role provides personal rewards for physicians as well as benefits for patients; it underlies and justifies the high value our society has placed on medicine.

On the other hand, directing our caregiving to the people we identify as sick or at risk of becoming sick may reinforce our propensity to categorize people as healthy or sick, as normal or abnormal. Labels such as obesity and hypertension focus attention on the tails of the distributions of risk factors and obscure the fact that there is no threshold that separates low risk from high risk. There is a danger that we will come to think of patients as constituting a separate population of "others," whose medical problems are unrelated to those in the broader community. In truth, the people who get sick from man-made diseases are not distinct subpopulations but merely larger or smaller fractions of the population as a whole (Law and Wald 2002). They are the "tip of the iceberg" and their prevalence cannot be understood without understanding the populations from which they come (Watt 1996).

An evolutionary view of disease would direct our attention to the populations or communities in which our patients live. Human populations evolved as geographically isolated populations. These populations have come to differ genetically because they were reproductively isolated but the genetic differences are secondary to the different environments in which the populations have lived. The genetic differences within human populations are greater than the differences among them (Rosenberg et al. 2002). The incidence of disease varies dramatically among human populations. These differences are due primarily to cultural, or man-made, rather than genetic differences among populations. The rapidly changing incidence of the chronic, noncommunicable diseases we have discussed is strong evidence that their prevalence is due primarily to man-made changes in the human environment.

As the epidemiologist Geoffrey Rose has emphasized, we need to be concerned with the health of populations as well as the health of individuals. As he noted, "The determinants of incidence are not necessarily the same as the causes of cases" (Rose 2001). While we need to investigate the causes of cases, which may be due in large part to genetic variation among individuals, we need to be sure that this does not divert or distract us from understanding and ameliorating the determinants of incidence, the cultural factors that have increased the incidence of diseases in modern populations. The cultural practices or traditions that affect the incidence of disease are transmitted from generation to generation and so can persist for long periods but they can also be rapidly modified or changed.

Rose argued that disease prevention requires a "population strategy" as well as an individual "high-risk strategy." There is ample precedent for population-based interventions, from vaccination to improvements in water supplies, food additives such as iodized salt and vitamin A- and D-fortified milk, anti-smoking campaigns, laws banning lead in gasoline and paint, and advances in automotive safety, that have dramatically improved our health and well-being. Reducing the burden of the diseases that afflict us today will require, among other things, population-wide changes in diet and physical activity, and reduction in socioeconomic disparities. It will be challenging for physicians or public health workers to effect these changes. It is unlikely that politicians who don't accept the theory of evolution will be persuaded by appeals to an evolutionary understanding of human beings and human populations. Population-based interventions are often opposed on the grounds that they conflict with other values, such as individual autonomy and free markets. It is more difficult to document the value of population-based interventions than of interventions targeted to individuals. And public health interventions must be carefully planned and implemented, because there is always a risk that they will be accompanied by unexpected and unwanted complications, the way HIV was inadvertently spread by health care workers in the early twentieth century (see Chapter 8; Pepin 2011). Nonetheless, as vaccination and the other interventions mentioned above demonstrate, population-based strategies of disease prevention can provide extraordinary benefits, and often at low cost. Helping us pay attention to the health of populations as well as the health of individuals is one more way in which evolutionary medicine will become an invaluable part of the broader field of medicine.

Glossary

Adaptation A trait that has spread in a population by natural selection because it enhanced the fitness of organisms in that population.

Allele One specific form or nucleotide sequence of a gene. Different alleles may or may not produce different phenotypic effects in the organisms that carry them.

Antagonistic pleiotropy The situation in which a gene has multiple phenotypic effects, which have opposing effects on fitness.

Cline A gradient in allele frequencies or phenotypes, usually with respect to latitude or altitude.

Demographic transition A model for the way populations change from having high birth and death rates to having low birth and death rates.

Diploid A cell or organism whose genome contains two sets of chromosomes. Distinct from haploid (one set) and polyploid (many sets).

Dominant An allele that produces the same phenotype in heterozygous and homozygous organisms.

Epigenetic Heritable changes in gene expression and in phenotype, either across generations or in somatic cell lineages, that are not dependent on changes in DNA sequence.

Epistasis Nonadditive interactions between two or more genes (or gene products) in determining a phenotypic trait.

Fitness The average reproductive success of organisms of a specific genotype, relative to the reproductive success of other organisms in the population.

Fixation The state in which one allele has replaced all other alleles at a genetic locus.

Founder effect The allele frequencies in a population that reflect the alleles that happened to be present in the founders of that population.

Frequency dependent selection The situation in which the fitness effects of an allele depend upon its frequency in the population.

Gene pool The set of genes shared by a population of interbreeding organisms.

Genetic drift Random changes in allele frequencies in a population.

Genotype The set of alleles at a single locus or in the genome of an organism.

Haplotype The set of alleles in a haploid genome, which has only one allele at each genetic locus.

Heterozygote advantage The situation in which a heterozygous organism has greater fitness than homozygous organisms. Because heterozygous organisms pass on both alleles to their offspring, heterozygote advantage maintains genetic polymorphisms.

Heterozygous Having different alleles at a genetic locus.

Hitchhiking The spread of an allele in a population because it is genetically linked to a beneficial allele.

Hominin Species on the lineage that led to the evolution of human beings, after this lineage diverged from the lineage that led to chimpanzees.

Homozygous Having the same alleles at a genetic locus.

Inclusive fitness An organism's own fitness plus its contribution to the fitness of its genetic relatives.

Microbiome The communities of microorganisms that normally inhabit our skin and body cavities.

Mutation-selection balance The process that maintains the frequency of deleterious alleles roughly constant because the rate at which they are formed by new mutation equals the rate at which they are lost as a result of natural selection.

Norm of reaction The range of phenotypes among organisms of a given genotype that develop in different environments.

Paleolithic A period of human history corresponding roughly in time to the Pleistocene epoch and characterized by the development and use of relatively simple stone tools.

Phenotype The anatomic, physiologic, and behavioral properties of an organism.

Pleiotropic A gene that affects more than one phenotypic trait.

Pleistocene A geologic period, now dated from roughly 2.6 million to 10–12 000 years ago, that was characterized by repeated glaciations and during which *Homo sapiens* arose.

Polymorphism The existence of a gene as two or more alleles within a population, each with frequencies too high to be due to mutation-selection balance.

Recessive An allele that produces a distinct phenotype only in homozygous organisms.

Replacement rate The fertility rate that would maintain a stationary (nongrowing) population, i.e. in which each adult woman would on average produce one daughter who herself survived to adulthood.

Sexual selection The component of natural selection that involves competition with other members of the same sex for access to mating partners, often through being chosen by members of the opposite sex.

Total fertility rate (TFR) The number of children a woman would have if her fertility corresponded to the current age-specific fertility rates in her society.

Zoonosis A disease that is transmitted from animals to humans.

References

Adrogue, H. J. and Madias, N. E. (2007). Sodium and potassium in the pathogenesis of hypertension. *New England Journal of Medicine*, **356**, 1966–78.

Allison, A. C. (1954). The distribution of the sickle-cell trait in East Africa and elsewhere, and its apparent relationship to the incidence of subtertian malaria. *Transactions of the Royal Society of Tropical Medicine and Hygiene*, **48**, 312–18.

Allison, A. C. (1964). Polymorphism and natural selection in human populations. *Cold Spring Harbor Symposia on Quantitative Biology*, **29**, 137–49.

Ames, B. N., Shigenaga, M. K., and Gold, L. S. (1993). DNA lesions, inducible DNA repair, and cell division: three key factors in mutagenesis and carcinogenesis. *Environmental Health Perspectives*, **101**, Suppl 5, 35–44.

Andersen, D. H. (1938). Cystic fibrosis of the pancreas and its relation to celiac disease: a clinical and pathological study. *American Journal of Diseases of Children*, **56**, 344–99.

Anderson, R. M. and May, R. M. (1991). *Infectious diseases of humans: dynamics and control*. Oxford University Press, Oxford.

Antonovics, J., Boots, M., Abbate, J., et al. (2011). Biology and evolution of sexual transmission. *Annals of the New York Academy of Sciences*, **1230**, 12–24.

Arbuthnott, J. (1710–1711). An argument for divine providence, taken from the constant regularity observ'd in the births of both sexes. *Philosophical Transactions of the Royal Society*, **27**, 186–90.

Aristotle (c.350 BCE). On youth and old age, on life and death, on breathing, <http://classics.mit.edu/Aristotle/youth_old.html>, accessed 30 September 2012.

Armelagos, G. J., Brown, P. J., and Turner, B. (2005). Evolutionary, historical and political economic perspectives on health and disease. *Social Science and Medicine*, **61**, 755–65.

Auricchio, S. and Maiuri, L. (1994). Cellular basis of adult-type hypolactasia. *Acta Paediatrica. Supplement*, **83**, 14–17.

Barakat, T. S., Jonkers, I., Monkhorst, K., et al. (2010). X-changing information on X inactivation. *Experimental Cell Research*, **316**, 679–87.

Barker, D. J. P. (2004). The developmental origins of well-being. *Philosophical Transactions of the Royal Society B: Biological Sciences*, **359**, 1359–66.

Barnes, K. C. (2011). Genetic studies of the etiology of asthma. *Proceedings of the American Thoracic Society*, **8**, 143–48.

Barnes, K. C., Grant, A. V., and Gao, P. (2005). A review of the genetic epidemiology of resistance to parasitic disease and atopic asthma: common variants for common phenotypes? *Current Opinion in Allergy and Clinical Immunology*, **5**, 379–85.

Bartolomei, M. S. and Ferguson-Smith, A. C. (2011). Mammalian genomic imprinting. *Cold Spring Harbor Perspectives in Biology*, **3**, doi: 10.1101/cshperspect.a002592.

Bateson, P. (2001). Fetal experience and good adult design. *International Journal of Epidemiology*, **30**, 928–34.

Bayless, T. M. and Rosensweig, N. S. (1966). A racial difference in incidence of lactase deficiency: a survey of milk intolerance and lactase deficiency in healthy adult males. *JAMA*, **197**, 968–72.

Bayless, T. M., Rothfeld, B., Massa, C., et al. (1975). Lactose and milk intolerance: clinical implications. *New England Journal of Medicine*, **292**, 1156–59.

Beltowski, J. (2001). Guanylin and related peptides. *Journal of Physiology and Pharmacology*, **52**, 351–75.

Beraldo, F. H. and Garcia, C. R. (2005). Products of tryptophan catabolism induce Ca²⁺ release and modulate the cell cycle of *Plasmodium falciparum* malaria parasites. *Journal of Pineal Research*, **39**, 224–30.

Bernard, C. (1957). *An introduction to the study of experimental medicine*, trans. H. C. Greene. Dover, New York.

Bersaglieri, T., Sabeti, P. C., Patterson, N., et al. (2004). Genetic signatures of strong recent positive selection at the lactase gene. *American Journal of Human Genetics*, **74**, 1111–20.

Bishop, N. A. and Guarente, L. (2007). Genetic links between diet and lifespan: shared mechanisms from yeast to humans. *Nature Reviews Genetics*, **8**, 835–44.

Bissell, M. J. and Hines, W. C. (2011). Why don't we get more cancer? A proposed role of the microenvironment in restraining cancer progression. *Nature Medicine*, **17**, 320–29.

Black, R. E., Allen, L. H., Bhutta, Z. A., et al. (2008). Maternal and child undernutrition: global and regional exposures and health consequences. *Lancet*, **371**, 243–60.

Blaustein, M. P., Leenen, F. H., Chen, L., et al. (2012). How NaCl raises blood pressure: a new paradigm for the pathogenesis of salt-dependent hypertension. *American Journal of Physiology – Heart and Circulatory Physiology*, **302**, H1031–49.

Bonner, J. T. (1993). *Life cycles: reflections of an evolutionary biologist*. Princeton University Press, Princeton, NJ.

Borgerhoff Mulder, M. (1998). The demographic transition: are we any closer to an evolutionary explanation? *Trends in Ecology and Evolution*, **13**, 266–70.

Bowlby, J. (1982). *Attachment*. Basic Books, New York.

Boyd, R. and Richerson, P. J. (1985). *Culture and the evolutionary process*. University of Chicago Press, Chicago.

Boyd, R., Richerson, P. J., and Henrich, J. (2011). The cultural niche: why social learning is essential for human adaptation. *Proceedings of the National Academy of Sciences, USA*, **108**, Suppl 2, 10 918–25.

Brotman, R. M. (2011). Vaginal microbiome and sexually transmitted infections: an epidemiologic perspective. *Journal of Clinical Investigation*, **121**, 4610–17.

Buss, L. W. (1987). *The evolution of individuality*. Princeton University Press, Princeton, NJ.

Byars, S. G., Ewbank, D., Govindaraju, D. R., et al. (2010). Natural selection in a contemporary human population. *Proceedings of the National Academy of Sciences, USA*, **107**, Suppl 1, 1787–92.

Bynum, W. F. (1983). Darwin and the doctors: evolution, diathesis, and germs in 19th-century Britain. *Gesnerus*, **40**, 43–53.

Bynum, W. F. (2008). *The history of medicine: a very short introduction*. Oxford University Press, Oxford.

Cairns, J. (1975). Mutation selection and the natural history of cancer. *Nature*, **255**, 197–200.

Carter, R. and Mendis, K. N. (2002). Evolutionary and historical aspects of the burden of malaria. *Clinical Microbiology Reviews*, **15**, 564–94.

Cavalli-Sforza, L. L., Menozzi, P., and Piazza, A. (1994). *The history and geography of human genes*. Princeton University Press, Princeton, NJ.

Central Intelligence Agency (2012). The world factbook, <https://www.cia.gov/library/publications/the-world-factbook/index.html>, accessed 30 September 2012.

Chandola, T., Ferrie, J., Sacker, A., et al. (2007). Social inequalities in self reported health in early old age: follow-up of prospective cohort study. *BMJ*, **334**, 990.

Charnov, E. L. (1993). *Life history invariants: some explorations of symmetry in evolutionary ecology*. Oxford University Press, Oxford.

Chisholm, J. S. (1999). *Death, hope and sex*. Cambridge University Press, Cambridge.

Cho, I. and Blaser, M. J. (2012). The human microbiome: at the interface of health and disease. *Nature Reviews Genetics*, **13**, 260–70.

Chow, Y. K., Hirsch, M. S., Merrill, D. P., et al. (1993). Use of evolutionary limitations of HIV-1 multidrug resistance to optimize therapy. *Nature*, **361**, 650–54.

Cleaveland, S., Laurenson, M. K., and Taylor, L. H. (2001). Diseases of humans and their domestic mammals: pathogen characteristics, host range and the risk of emergence. *Philosophical Transactions of the Royal Society B: Biological Sciences*, **356**, 991–99.

Coale, A. J. (1974). The history of the human population. *Scientific American*, **231**, 41–51.

Cohen, J. E. (1995). *How many people can the earth support?* W. W. Norton, New York.

Cohuet, A., Harris, C., Robert, V., et al. (2010). Evolutionary forces on *Anopheles*: what makes a malaria vector? *Trends in Parasitology*, **26**, 130–36.

Cooke, A. (2012). Parasitic worms and inflammatory disease. *Current Opinion in Rheumatology*, **24**, 394–400.

Cordain, L., Eaton, S. B., Sebastian, A., et al. (2005). Origins and evolution of the Western diet: health implications for the 21st century. *American Journal of Clinical Nutrition*, **81**, 341–54.

Correale, J. and Farez, M. F. (2011). The impact of environmental infections (parasites) on MS activity. *Multiple Sclerosis Journal*, **17**, 1162–69.

Crimmins, E. M., Kim, J. K., and Seeman, T. E. (2009). Poverty and biological risk: the earlier "aging" of the poor. *Journals of Gerontology. Series A, Biological Sciences and Medical Sciences*, **64**, 286–92.

Cronin, H. (1991). *The ant and the peacock: altruism and sexual selection from Darwin to today.* Cambridge University Press, New York.

Crosby, A. W. (1971). *The Columbian exchange: biological and cultural consequences of 1492.* Greenwood, Westport, CT.

Crow, J. F. (1997). The high spontaneous mutation rate: is it a health risk? *Proceedings of the National Academy of Sciences, USA*, **94**, 8380–86.

Cuatrecasas, P., Lockwood, D. H., and Caldwell, J. R. (1965). Lactase deficiency in the adult. A common occurrence. *Lancet*, **1**, 14–18.

Curtis, V. (2011). Why disgust matters. *Philosophical Transactions of the Royal Society B: Biological Sciences*, **366**, 3478–90.

Cutler, R. G. (1991). Antioxidants and aging. *American Journal of Clinical Nutrition*, **53**, 373S–79S.

Cutting, G. R. (2010). Modifier genes in Mendelian disorders: the example of cystic fibrosis. *Annals of the New York Academy of Sciences*, **1214**, 57–69.

Cystic Fibrosis Mutation Database (no date). <http://www.genet.sickkids.on.ca/cftr>, accessed 30 September 2012.

Darwin, C. (1859). *On the origin of species by means of natural selection.* John Murray, London.

Darwin, C. (1871). *The descent of man, and selection in relation to sex.* John Murray, London.

Darwin, C. (1883). *The variation of animals and plants under domestication* 2nd edn. D. Appleton and Company, New York.

David, M. Z. and Daum, R. S. (2010). Community-associated methicillin-resistant *Staphylococcus aureus*: epidemiology and clinical consequences of an emerging epidemic. *Clinical Microbiology Reviews*, **23**, 616–87.

Dawkins, R. (1995). God's utility function. *Scientific American*, **273**, 80–85.

De Braekeleer, M., Daigneault, J., Allard, C., et al. (1996). Genealogy and geographical distribution of CFTR mutations in Saguenay Lac-Saint-Jean (Quebec, Canada). *Annals of Human Biology*, **23**, 345–52.

Di Rienzo, A. and Hudson, R. R. (2005). An evolutionary framework for common diseases: the ancestral-susceptibility model. *Trends in Genetics*, **21**, 596–601.

di Sant'Agnese, P. A., Darling, R. C., Perera, G. A., et al. (1953). Abnormal electrolyte composition of sweat in cystic fibrosis of the pancreas. *Pediatrics*, **12**, 549–63.

Diamond, J. M. (1992). *The third chimpanzee: the evolution and future of the human animal.* HarperCollins, New York.

Diamond, J. M. (1997). *Guns, germs, and steel: the fates of human societies.* W. W. Norton, New York.

Distante, S., Robson, K. J., Graham-Campbell, J., et al. (2004). The origin and spread of the HFE-C282Y haemochromatosis mutation. *Human Genetics*, **115**, 269–79.

Doblhammer, G. and Oeppen, J. (2003). Reproduction and longevity among the British peerage: the effect of frailty and health selection. *Proceedings of the Royal Society B: Biological Sciences*, **270**, 1541–47.

Drake, A. J., McPherson, R. C., Godfrey, K. M., et al. (2012). An unbalanced maternal diet in pregnancy associates with offspring epigenetic changes in genes controlling glucocorticoid action and fetal growth. *Clinical Endocrinology*, Epub 31 May 2012.

Dubos, R. (1942). Microbiology. *Annual Review of Biochemistry*, **11**, 659–78.

Durham, W. H. (1991). *Coevolution*. Stanford University Press, Stanford, CA.

Eaton, S. B., Cordain, L., and Sparling, P. B. (2009). Evolution, body composition, insulin receptor competition, and insulin resistance. *Preventive Medicine*, **49**, 283–85.

Eaton, S. B. and Konner, M. (1985). Paleolithic nutrition. A consideration of its nature and current implications. *New England Journal of Medicine*, **312**, 283–89.

Eaton, S. B., Pike, M. C., Short, R. V., et al. (1994). Women's reproductive cancers in evolutionary context. *Quarterly Review of Biology*, **69**, 353–67.

Eliot, T. S. (1943). East coker. *Four quartets*. Harcourt, Brace and Co., New York.

Ellison, P. T. (2001). *On fertile ground*. Harvard University Press, Cambridge, MA.

Enattah, N. S., Sahi, T., Savilahti, E., et al. (2002). Identification of a variant associated with adult-type hypolactasia. *Nature Genetics*, **30**, 233–37.

Esteller, M. (2011). Non-coding RNAs in human disease. *Nature Reviews Genetics*, **12**, 861–74.

Evershed, R. P., Payne, S., Sherratt, A. G., et al. (2008). Earliest date for milk use in the Near East and southeastern Europe linked to cattle herding. *Nature*, **455**, 528–31.

Ewald, P. W. (1980). Evolutionary biology and the treatment of signs and symptoms of infectious disease. *Journal of Theoretical Biology*, **86**, 169–76.

Ewald, P. W. (1994). *Evolution of infectious disease*. Oxford University Press, Oxford.

Ewald, P. W. and Swain Ewald, H. A. (2012). Infection, mutation, and cancer evolution. *Journal of Molecular Medicine*, **90**, 535–41.

Ezzati, M., Lopez, A. D., Rodgers, A., et al. (2002). Selected major risk factors and global and regional burden of disease. *Lancet*, **360**, 1347–60.

Fearon, E. R. and Vogelstein, B. (1990). A genetic model for colorectal tumorigenesis. *Cell*, **61**, 759–67.

Fell, D. (1997). *Understanding the control of metabolism*. Portland Press, London.

Fenner, F. (1983). The Florey lecture, 1983. Biological control, as exemplified by smallpox eradication and myxomatosis. *Proceedings of the Royal Society B: Biological Sciences*, **218**, 259–85.

Fisher, R. A. (1930). *The genetical theory of natural selection* 2nd rev. edn. Oxford University Press, Oxford.

Flatz, G. and Rotthauwe, H. W. (1973). Lactose nutrition and natural selection. *Lancet*, **2**, 76–77.

Fleming, J. O., Isaak, A., Lee, J. E., et al. (2011). Probiotic helminth administration in relapsing-remitting multiple sclerosis: a phase 1 study. *Multiple Sclerosis Journal*, **17**, 743–54.

Flexner, A. (1910). *Medical education in the United States and Canada: a report to the Carnegie Foundation for the Advancement of Teaching*. The Carnegie Foundation, New York.

Fogel, R. W. (2003). Secular trends in physiological capital: implications for equity in health care. *Perspectives in Biology and Medicine*, **46**, S24–S38.

Forsman, A. and Weiss, R. A. (2008). Why is HIV a pathogen? *Trends in Microbiology*, **16**, 555–60.

Fraga, C. G., Shigenaga, M. K., Park, J. W., et al. (1990). Oxidative damage to DNA during aging: 8-hydroxy-2'-deoxyguanosine in rat organ DNA and urine. *Proceedings of the National Academy of Sciences, USA*, **87**, 4533–37.

Frank, N. Y., Schatton, T., and Frank, M. H. (2010). The therapeutic promise of the cancer stem cell concept. *Journal of Clinical Investigation*, **120**, 41–50.

Fraser, C., Riley, S., Anderson, R. M., et al. (2004). Factors that make an infectious disease outbreak controllable. *Proceedings of the National Academy of Sciences, USA*, **101**, 6146–51.

Frassetto, L., Morris, R. C., Jr., Sellmeyer, D. E., et al. (2001). Diet, evolution and aging—the pathophysiologic effects of the post-agricultural inversion of the potassium-to-sodium and base-to-chloride ratios in the human diet. *European Journal of Nutrition*, **40**, 200–13.

Frisch, R. E. (2002). *Female fertility and the body fat connection.* University of Chicago Press, Chicago.

Garrod, A. E. (1909). *Inborn errors of metabolism.* Henry Frowde and Hodder & Stoughton, London.

Gerbault, P., Liebert, A., Itan, Y., et al. (2011). Evolution of lactase persistence: an example of human niche construction. *Philosophical Transactions of the Royal Society B: Biological Sciences,* **366**, 863–77.

Gerlinger, M., Rowan, A. J., Horswell, S., et al. (2012). Intratumor heterogeneity and branched evolution revealed by multiregion sequencing. *New England Journal of Medicine,* **366**, 883–92.

Giacani, L., Molini, B. J., Kim, E. Y., et al. (2010). Antigenic variation in *Treponema pallidum*: TprK sequence diversity accumulates in response to immune pressure during experimental syphilis. *Journal of Immunology,* **184**, 3822–29.

Gilbert, S. F. (2001). Ecological developmental biology: developmental biology meets the real world. *Developmental Biology,* **233**, 1–12.

Gluckman, P. D., Beedle, A. S., and Hanson, M. A. (2009). *Principles of evolutionary medicine.* Oxford University Press, Oxford.

Gluckman, P. D. and Hanson, M. A. (2006). *Mismatch: why our world no longer fits our bodies.* Oxford University Press, Oxford.

Gluckman, P. D., Hanson, M. A., and Beedle, A. S. (2007). Early life events and their consequences for later disease: a life history and evolutionary perspective. *American Journal of Human Biology,* **19**, 1–19.

Gluckman, P. D., Hanson, M. A., and Butklijas, T. (2010). A conceptual framework for the developmental origins of health and disease. *Journal of Developmental Origins of Health and Disease,* **1**, 6–18.

Goldenberg, M. J. (2009). Iconoclast or creed? Objectivism, pragmatism, and the hierarchy of evidence. *Perspectives in Biology and Medicine,* **52**, 168–87.

Goodman, A., Koupil, I., and Lawson, D. W. (2012). Low fertility increases descendant socioeconomic position but reduces long-term fitness in a modern post-industrial society. *Proceedings of the Royal Society B: Biological Sciences,* **279**, 4342–51.

Greaves, M. (2002). Cancer causation: the Darwinian downside of past success? *Lancet Oncology,* **3**, 244–51.

Greaves, M. and Maley, C. C. (2012). Clonal evolution in cancer. *Nature,* **481**, 306–13.

Grunewald, T. G., Herbst, S. M., Heinze, J., et al. (2011). Understanding tumor heterogeneity as functional compartments—superorganisms revisited. *Journal of Translational Medicine,* **9**, 79.

Haldane, J. B. S. (1949a). Disease and evolution. *La Ricerca Scientifica,* **19**, 68–76.

Haldane, J. B. S. (1949b). The rate of mutation of human genes. *Hereditas,* **35**, 267–73.

Hales, C. N. and Barker, D. J. (2001). The thrifty phenotype hypothesis. *British Medical Bulletin,* **60**, 5–20.

Hamilton, B. E., Martin, J. A., and Ventura, S. J. (2010). Births: preliminary data for 2009. *National Vital Statistics Reports,* **59**, National Center for Health Statistics, Centers for Disease Control and Prevention.

Hamilton, J. B. and Mestler, G. E. (1969). Mortality and survival: comparison of eunuchs with intact men and women in a mentally retarded population. *Journal of Gerontology,* **24**, 395–411.

Hamilton, W. D. (1964). The genetical evolution of social behaviour. I. *Journal of Theoretical Biology,* **7**, 1–16.

Hamilton, W. D. (1966). The moulding of senescence by natural selection. *Journal of Theoretical Biology,* **12**, 12–45.

Hanahan, D. and Weinberg, R. A. (2000). The hallmarks of cancer. *Cell,* **100**, 57–70.

Hanahan, D. and Weinberg, R. A. (2011). Hallmarks of cancer: the next generation. *Cell,* **144**, 646–74.

Hancock, A. M. and Di Rienzo, A. (2008). Detecting the genetic signature of natural selection in human populations: models, methods, and data. *Annual Review of Anthropology,* **37**, 197–217.

Hardin, G. (1960). The competitive exclusion principle. *Science,* **131**, 1292–97.

Harding, C., Pompei, F., and Wilson, R. (2012). Peak and decline in cancer incidence, mortality, and prevalence at old ages. *Cancer,* **118**, 1371–86.

Harper, K. N., Ocampo, P. S., Steiner, B. M., et al. (2008). On the origin of the treponematoses: a phylogenetic approach. *PLoS Neglected Tropical Diseases,* **2**, e148.

Hart, R. W. and Setlow, R. B. (1974). Correlation between deoxyribonucleic acid excision-repair and lifespan in a number of mammalian species. *Proceedings of the National Academy of Sciences, USA*, **71**, 2169–73.

Hartl, D. L. and Clark, A. G. (2006). *Principles of population genetics* 4th edn. Sinauer Associates, Sunderland, MA.

Hawkes, K. (2004). Human longevity: the grandmother effect. *Nature*, **428**, 128–29.

Hayflick, L. (1980). Recent advances in the cell biology of aging. *Mechanisms of Ageing and Development*, **14**, 59–79.

He, F. J. and MacGregor, G. A. (2009). A comprehensive review on salt and health and current experience of worldwide salt reduction programmes. *Journal of Human Hypertension*, **23**, 363–84.

Hedrick, P. (2004). Estimation of relative fitnesses from relative risk data and the predicted future of haemoglobin alleles S and C. *Journal of Evolutionary Biology*, **17**, 221–24.

Held, L. I., Jr. (2009). *Quirks of human anatomy: an evo-devo look at the human body*. Cambridge University Press, Cambridge.

Hershkovitz, D., Burbea, Z., Skorecki, K., et al. (2007). Fetal programming of adult kidney disease: cellular and molecular mechanisms. *Clinical Journal of the American Society of Nephrology*, **2**, 334–42.

Hesslow, G. (1984). What is a genetic disease? On the relative importance of causes. In L. Nordenfelt and B. I. B. Lindahl (eds.), *Health, disease and causal explanations in medicine*, 183–93. D. Reidel, Dordrecht.

Hill, K. (1993). Life history theory and evolutionary anthropology. *Evolutionary Anthropology*, **2**, 78–88.

Hobbs, F. and Stoops, N. (2002). *Demographic trends in the 20th century*, U.S. Census Bureau, Census 2000 Special Reports. U.S. Government Printing Office, Washington, DC.

Hodgkin, P. (1985). Medicine is war: and other medical metaphors. *British Medical Journal (Clinical Research Ed.)*, **291**, 1820–21.

Holden, C. and Mace, R. (1997). Phylogenetic analysis of the evolution of lactose digestion in adults. *Human Biology*, **69**, 605–28.

Hollingsworth, T. D., Anderson, R. M., and Fraser, C. (2008). HIV-1 transmission, by stage of infection. *Journal of Infectious Diseases*, **198**, 687–93.

Huber, M., Knottnerus, J. A., Green, L., et al. (2011). How should we define health? *BMJ*, **343**, d4163.

Hudson, E. H. (1965). Treponematosis in perspective. *Bulletin of the World Health Organization*, **32**, 735–48.

Hunt, D. M., Carvalho, L. S., Cowing, J. A., et al. (2009). Evolution and spectral tuning of visual pigments in birds and mammals. *Philosophical Transactions of the Royal Society B: Biological Sciences*, **364**, 2941–55.

Huxley, J. S. (1942). *Evolution: the modern synthesis*. G. Allen & Unwin, London.

Intersalt Cooperative Research Group (1988). Intersalt: an international study of electrolyte excretion and blood pressure. Results for 24 hour urinary sodium and potassium excretion. *BMJ*, **297**, 319–28.

Itan, Y., Jones, B. L., Ingram, C. J., et al. (2010). A worldwide correlation of lactase persistence phenotype and genotypes. *BMC Evolutionary Biology*, **10**, 36.

Jablonka, E. and Lamb, M. J. (2005). *Evolution in four dimensions*. MIT Press, Cambridge, MA.

Jablonka, E. and Raz, G. (2009). Transgenerational epigenetic inheritance: prevalence, mechanisms, and implications for the study of heredity and evolution. *Quarterly Review of Biology*, **84**, 131–76.

Jablonski, N. G. and Chaplin, G. (2010). Human skin pigmentation as an adaptation to UV radiation. *Proceedings of the National Academy of Sciences, USA*, **107**, Suppl 2, 8962–68.

Jackson, J. H. (1884). The Croonian Lectures on evolution and dissolution of the nervous system. Lecture 1. *British Medical Journal*, **1**, 591–93.

Jacob, F. (1976). *The logic of life*. Vintage, New York.

Jones, S., Chen, W. D., Parmigiani, G., et al. (2008). Comparative lesion sequencing provides insights into tumor evolution. *Proceedings of the National Academy of Sciences, USA*, **105**, 4283–88.

Jordan, I. K., Kota, K. C., Cui, G., et al. (2008). Evolutionary and functional divergence between the cystic fibrosis transmembrane conductance regulator and related ATP-binding cassette transporters. *Proceedings of the National Academy of Sciences, USA*, **105**, 18865–70.

Jorde, L. B. and Lathrop, G. M. (1988). A test of the heterozygote-advantage hypothesis in cystic fibrosis carriers. *American Journal of Human Genetics*, **42**, 808–15.

Kacser, H. and Burns, J. A. (1981). The molecular basis of dominance. *Genetics*, **97**, 639–66.

Kallus, S. J. and Brandt, L. J. (2012). The intestinal microbiota and obesity. *Journal of Clinical Gastroenterology*, **46**, 16–24.

Kaplan, H., Hill, K., Lancaster, J., et al. (2000). A theory of human life history evolution: diet, intelligence, and longevity. *Evolutionary Anthropology*, **9**, 156–84.

Kaplan, H., Lancaster, J., and Robson, A. (2003). Embodied capital and the evolutionary economics of the human life span. In J. R. Carey and S. Tuljapurkar (eds.), *Life span: evolutionary, ecological, and demographic perspectives*, 152–82. Population Council, New York.

Kaplan, H., Lancaster, J. B., Tucker, W. T., et al. (2002). Evolutionary approach to below replacement fertility. *American Journal of Human Biology*, **14**, 233–56.

Keinan, A. and Clark, A. G. (2012). Recent explosive human population growth has resulted in an excess of rare genetic variants. *Science*, **336**, 740–43.

Kevles, D. J. (1995). *In the name of eugenics*. Harvard University Press, Cambridge, MA.

Kirkwood, T. B. L. (1996). Human senescence. *Bioessays*, **18**, 1009–16.

Kirkwood, T. B. L. and Austad, S. N. (2000). Why do we age? *Nature*, **408**, 233–38.

Klein, R. G. (2009). *The human career: human biological and cultural origins* 3rd edn. University of Chicago Press, Chicago.

Kluger, M. J., Kozak, W., Conn, C. A., et al. (1996). The adaptive value of fever. *Infectious Disease Clinics of North America*, **10**, 1–20.

Konner, M. and Eaton, S. B. (2010). Paleolithic nutrition: twenty-five years later. *Nutrition in Clinical Practice*, **25**, 594–602.

Kopic, S. and Geibel, J. P. (2010). Toxin mediated diarrhea in the 21st century: the pathophysiology of intestinal ion transport in the course of ETEC, *V. cholerae* and Rotavirus infection. *Toxins (Basel)*, **2**, 2132–57.

Kosova, G., Pickrell, J. K., Kelley, J. L., et al. (2010). The CFTR Met 470 allele is associated with lower birth rates in fertile men from a population isolate. *PLoS Genetics*, **6**, e1000974.

Kretchmer, N. (1972). Lactose and lactase. *Scientific American*, **227**, 71–78.

Kruger, D. J. and Nesse, R. M. (2004). Sexual selection and the male:female mortality ratio. *Evolutionary Psychology*, **2**, 66–85.

Kurlansky, M. (2002). *Salt*. Penguin Books, New York.

Kwiatkowski, D. P. (2005). How malaria has affected the human genome and what human genetics can teach us about malaria. *American Journal of Human Genetics*, **77**, 171–92.

Laberge, A. M., Michaud, J., Richter, A., et al. (2005). Population history and its impact on medical genetics in Quebec. *Clinical Genetics*, **68**, 287–301.

LaFond, R. E. and Lukehart, S. A. (2006). Biological basis for syphilis. *Clinical Microbiology Reviews*, **19**, 29–49.

Lakoff, G. and Johnson, M. (2003). *Metaphors we live by*. University of Chicago Press, Chicago.

Laland, K. N., Sterelny, K., Odling-Smee, F. J., et al. (2011). Cause and effect in biology revisited: is Mayr's proximate-ultimate dichotomy still useful? *Science*, **334**, 1512–16.

Lambert, P. M. (2009). Health versus fitness: competing themes in the origins and spread of agriculture? *Current Anthropology*, **50**, 603–8.

Landry, D. J. and Forrest, J. D. (1995). How old are U.S. fathers? *Family Planning Perspectives*, **27**, 159–61, 65.

Law, M. R. and Wald, N. J. (2002). Risk factor thresholds: their existence under scrutiny. *BMJ*, **324**, 1570–76.

Lefevre, C. M., Sharp, J. A., and Nicholas, K. R. (2010). Evolution of lactation: ancient origin and extreme adaptations of the lactation system. *Annual Review of Genomics and Human Genetics*, **11**, 219–38.

Leigh, S. R. (2001). Evolution of human growth. *Evolutionary Anthropology*, **10**, 223–36.

Leroi, A. M., Koufopanou, V., and Burt, A. (2003). Cancer selection. *Nature Reviews Cancer*, **3**, 226–31.

Levin, B. R. and Bull, J. J. (1994). Short-sighted evolution and the virulence of pathogenic microorganisms. *Trends in Microbiology*, **2**, 76–81.

Levy, S. B. and Marshall, B. (2004). Antibacterial resistance worldwide: causes, challenges and responses. *Nature Medicine*, **10**, S122–29.

Lewontin, R. C. (1970). The units of selection. *Annual Review of Ecology and Systematics*, **1**, 1–18.

Lewontin, R. C. (1990). Darwin and Marx. *New York Review of Books*, **37**, 6 December.

Lieberman, M. and Lieberman, D. (1978). Lactase deficiency: a genetic mechanism which regulates the time of weaning. *American Naturalist*, **112**, 625–27.

Lockhart, A. B., Thrall, P. H., and Antonovics, J. (1996). Sexually transmitted diseases in animals: ecological and evolutionary implications. *Biological Reviews of the Cambridge Philosophical Society*, **71**, 415–71.

Loeb, L. A. (2011). Human cancers express mutator phenotypes: origin, consequences and targeting. *Nature Reviews Cancer*, **11**, 450–57.

Loftus, R. T., MacHugh, D. E., Bradley, D. G., et al. (1994). Evidence for two independent domestications of cattle. *Proceedings of the National Academy of Sciences, USA*, **91**, 2757–61.

Lopez, A. D. and Mathers, C. D. (2006). Measuring the global burden of disease and epidemiological transitions: 2002-2030. *Annals of Tropical Medicine and Parasitology*, **100**, 481–99.

Lustig, R. H., Schmidt, L. A., and Brindis, C. D. (2012). Public health: the toxic truth about sugar. *Nature*, **482**, 27–9.

Luzzatto, L., Usanga, F. A., and Reddy, S. (1969). Glucose-6-phosphate dehydrogenase deficient red cells: resistance to infection by malarial parasites. *Science*, **164**, 839–42.

Lynch, M. (2010). Rate, molecular spectrum, and consequences of human mutation. *Proceedings of the National Academy of Sciences, USA*, **107**, 961–68.

Mackenbach, J. P. and Howden-Chapman, P. (2003). New perspectives on socioeconomic inequalities in health. *Perspectives in Biology and Medicine*, **46**, 428–44.

Macklon, N. S., Geraedts, J. P., and Fauser, B. C. (2002). Conception to ongoing pregnancy: the 'black box' of early pregnancy loss. *Human Reproduction Update*, **8**, 333–43.

MacLean, P. D. (1990). *The triune brain in evolution*. Plenum Press, New York.

Malthus, T. R. (1798). *An essay on the principle of population*. J. Johnson, London.

Mantei, N., Villa, M., Enzler, T., et al. (1988). Complete primary structure of human and rabbit lactase-phlorizin hydrolase: implications for biosynthesis, membrane anchoring and evolution of the enzyme. *EMBO Journal*, **7**, 2705–13.

Marmot, M. G. (2004). *The status syndrome*. Times Books, New York.

Marshall, W. S. and Singer, T. D. (2002). Cystic fibrosis transmembrane conductance regulator in teleost fish. *Biochimica et Biophysica Acta*, **1566**, 16–27.

Martin, M. J., Rayner, J. C., Gagneux, P., et al. (2005). Evolution of human-chimpanzee differences in malaria susceptibility: relationship to human genetic loss of N-glycolylneuraminic acid. *Proceedings of the National Academy of Sciences, USA*, **102**, 12 819–24.

Marx, K. (1862). Letter to Friedrich Engels, 18 June, <http://www.marxists.org/archive/marx/works/1862/letters/62_06_18.htm>, accessed 30 September 2012.

Mason, P. (2011). *Medical neurobiology*. Oxford University Press, New York.

Maynard Smith, J. (1978). *The evolution of sex*. Cambridge University Press, Cambridge.

Mayr, E. (1964). Introduction. *Charles Darwin. On the origin of species*, vii–xxvii. Harvard University Press, Cambridge, MA.

Mayr, E. (1988a). Cause and effect in biology. *Toward a new philosophy of biology*, 24–37. Harvard University Press, Cambridge, MA. Originally published in *Science*, **134**, 1501–6, 1961.

Mayr, E. (1988b). The species category. *Toward a new philosophy of biology*, 315–34. Harvard University Press, Cambridge, MA.

McCracken, R. D. (1971). Lactase deficiency: an example of dietary evolution. *Current Anthropology*, **12**, 479–517.

McKeown, T. (1988). *The origins of human disease*. Blackwell, Oxford.

Medawar, P. B. (1952). *An unsolved problem of biology*. H. K. Lewis, London.

Mendel, G. (1865). Experiments in plant hybridization, <http://www.esp.org/foundations/genetics/classical/gm-65.pdf>, accessed 30 September 2012.

Merlo, L. M., Pepper, J. W., Reid, B. J., et al. (2006). Cancer as an evolutionary and ecological process. *Nature Reviews Cancer*, **6**, 924–35.

Min, K.-J., Lee, C.-K., and Park, H.-N. (2012). The lifespan of Korean eunuchs. *Current Biology*, **22**, R792–R93.

Monod, J. (1971). *Chance and necessity*. Knopf, New York.

Moss, S. F. and Blaser, M. J. (2005). Mechanisms of disease: inflammation and the origins of cancer. *Nature Clinical Practice Oncology*, **2**, 90–97.

Mukherjee, S. (2010). *The emperor of all maladies: a biography of cancer*. Scribner, New York.

Muller, V., Maggiolo, F., Suter, F., et al. (2009). Increasing clinical virulence in two decades of the Italian HIV epidemic. *PLoS Pathogens*, **5**, e1000454.

Mulligan, C. J., Norris, S. J., and Lukehart, S. A. (2008). Molecular studies in *Treponema pallidum* evolution: toward clarity? *PLoS Neglected Tropical Diseases*, **2**, e184.

Murray, C. J., Rosenfeld, L. C., Lim, S. S., et al. (2012). Global malaria mortality between 1980 and 2010: a systematic analysis. *Lancet*, **379**, 413–31.

Nahmias, A. and Danielsson, D. (2011). Introduction to *The evolution of infectious agents in relation to sex*. *Annals of the New York Academy of Sciences*, **1230**, xiii–xix.

National Center for Health Statistics (2012). Health, United States, 2011: with special feature on socioeconomic status and health, <http://www.cdc.gov/nchs/data/hus/hus11.pdf>, accessed 30 September 2012.

Navin, N. and Hicks, J. (2011). Future medical applications of single-cell sequencing in cancer. *Genome Medicine*, **3**, 31.

Neel, J. V. (1962). Diabetes mellitus: a 'thrifty' genotype rendered detrimental by 'progress'? *American Journal of Human Genetics*, **14**, 353–62.

Nesse, R. M. (2005). Maladaptation and natural selection. *Quarterly Review of Biology*, **80**, 62–70.

Nesse, R. M. and Williams, G. C. (1994). *Why we get sick: the new science of Darwinian medicine*. Times Books, New York.

Nettle, D. (2011). Flexibility in reproductive timing in human females: integrating ultimate and proximate explanations. *Philosophical Transactions of the Royal Society B: Biological Sciences*, **366**, 357–65.

Nowell, P. C. (1976). The clonal evolution of tumor cell populations. *Science*, **194**, 23–28.

Nunney, L. (1999). Lineage selection and the evolution of multistage carcinogenesis. *Proceedings of the Royal Society B: Biological Sciences*, **266**, 493–98.

Odling-Smee, F. J., Laland, K. N., and Feldman, M. W. (2003). *Niche construction: the neglected process in evolution*. Princeton University Press, Princeton, NJ.

Office for National Statistics (2011). Trends in life expectancy by the National Statistics socio-economic classification 1982–2006, <http://www.ons.gov.uk/ons/taxonomy/index.html?nscl=Health+Inequalities>, accessed 30 September 2012.

Oleksyk, T. K., Smith, M. W., and O'Brien, S. J. (2010). Genome-wide scans for footprints of natural selection. *Philosophical Transactions of the Royal Society B: Biological Sciences*, **365**, 185–205.

OMIM (no date). Online Mendelian Inheritance in Man, <http://www.ncbi.nlm.nih.gov/omim/>, accessed 30 September 2012.

Omran, A. R. (1971). The epidemiologic transition. A theory of the epidemiology of population change. *Milbank Memorial Fund Quarterly*, **49**, 509–38.

Paget, S. (1889). The distribution of secondary growths in cancer of the breast. *Lancet*, **133**, 571–73.

Penman, B. S., Pybus, O. G., Weatherall, D. J., et al. (2009). Epistatic interactions between genetic disorders of hemoglobin can explain why the sickle-cell gene is uncommon in the Mediterranean. *Proceedings of the National Academy of Sciences, USA*, **106**, 21242–46.

Pepin, J. (2011). *The origins of AIDS*. Cambridge University Press, Cambridge.

Perlman, R. L. (2005). Why disease persists: an evolutionary nosology. *Medicine, Health Care and Philosophy*, **8**, 343–50.

Perlman, R. L. (2008). Socioeconomic inequalities in ageing and health. *Lancet*, **372**, S34–S39.

Perlman, R. L. (2009). Life histories of pathogen populations. *International Journal of Infectious Diseases*, **13**, 121–24.

Peto, R., Darby, S., Deo, H., et al. (2000). Smoking, smoking cessation, and lung cancer in the UK since 1950: combination of national statistics with two case-control studies. *BMJ*, **321**, 323–29.

Phillips, D. I. (2007). Programming of the stress response: a fundamental mechanism underlying the long-term effects of the fetal environment? *Journal of Internal Medicine*, **261**, 453–60.

Pier, G. B. (2002). CFTR mutations and host susceptibility to *Pseudomonas aeruginosa* lung infection. *Current Opinion in Microbiology*, **5**, 81–86.

Pier, G. B., Grout, M., Zaidi, T., et al. (1998). *Salmonella typhi* uses CFTR to enter intestinal epithelial cells. *Nature*, **393**, 79–82.

Pollan, M. (2008). *In defense of food: an eater's manifesto*. Penguin, New York.

Pompei, F., Ciminelli, B. M., Bombieri, C., et al. (2006). Haplotype block structure study of the CFTR gene. Most variants are associated with the M470 allele in several European populations. *European Journal of Human Genetics*, **14**, 85–93.

Poolman, E. M. and Galvani, A. P. (2007). Evaluating candidate agents of selective pressure for cystic fibrosis. *Journal of the Royal Society Interface*, **4**, 91–98.

Poropatich, K. and Sullivan, D. J., Jr. (2011). Human immunodeficiency virus type 1 long-term non-progressors: the viral, genetic and immunological basis for disease non-progression. *Journal of General Virology*, **92**, 247–68.

Porter, R. (1998). *The greatest benefit to mankind: a medical history of humanity*. W.W. Norton, New York.

Powell, A., Shennan, S., and Thomas, M. G. (2009). Late Pleistocene demography and the appearance of modern human behavior. *Science*, **324**, 1298–301.

Profet, M. (1993). Menstruation as a defense against pathogens transported by sperm. *Quarterly Review of Biology*, **68**, 335–86.

Quétel, C. (1990). *History of syphilis*, trans. J. Braddock and B. Pike. Johns Hopkins University Press, Baltimore, MD.

Quinton, P. M. (1983). Chloride impermeability in cystic fibrosis. *Nature*, **301**, 421–22.

Rasinpera, H., Kuokkanen, M., Kolho, K. L., et al. (2005). Transcriptional downregulation of the lactase (LCT) gene during childhood. *Gut*, **54**, 1660–61.

Rasinpera, H., Savilahti, E., Enattah, N. S., et al. (2004). A genetic test which can be used to diagnose adult-type hypolactasia in children. *Gut*, **53**, 1571–76.

Reich, D. E. and Lander, E. S. (2001). On the allelic spectrum of human disease. *Trends in Genetics*, **17**, 502–10.

Rich, S. M., Leendertz, F. H., Xu, G., et al. (2009). The origin of malignant malaria. *Proceedings of the National Academy of Sciences, USA*, **106**, 14902–7.

Richards, A. L. (1997). Tumour necrosis factor and associated cytokines in the host's response to malaria. *International Journal for Parasitology*, **27**, 1251–63.

Richerson, P. J. and Boyd, R. (2005). *Not by genes alone: how culture transformed human evolution*. University of Chicago Press, Chicago.

Richman, D. D., Little, S. J., Smith, D. M., et al. (2004). HIV evolution and escape. *Transactions of the American Clinical and Climatological Association*, **115**, 289–303.

Riordan, J. R., Rommens, J. M., Kerem, B.-s., et al. (1989). Identification of the cystic fibrosis gene: cloning and characterization of the complementary DNA. *Science*, **245**, 1066–73.

Robson, S. L., van Schaik, C. P., and Hawkes, K. (2006). The derived features of human life history. In K. Hawkes and R. R. Paine (eds.), *The evolution of human life history*, 17–44. School of American Research Press, Santa Fe, NM.

Romeo, G., Devoto, M., and Galietta, L. J. (1989). Why is the cystic fibrosis gene so frequent? *Human Genetics*, **84**, 1–5.

Rook, G. A. (2012). Hygiene hypothesis and autoimmune diseases. *Clinical Reviews in Allergy and Immunology*, **42**, 5–15.

Rose, G. (2001). Sick individuals and sick populations. *International Journal of Epidemiology*, **30**, 427–32; discussion 33–34. Originally published in *International Journal of Epidemiology*, **14**, 32–38, 1985.

Rose, M. R. (1991). *Evolutionary biology of aging*. Oxford University Press, New York.

Rose, M. R. and Mueller, L. D. (1998). Evolution of human lifespan: past, future, and present. *American Journal of Human Biology*, **10**, 409–20.

Rose, M. R., Rauser, C. L., Mueller, L. D., et al. (2006). A revolution for aging research. *Biogerontology*, **7**, 269–77.

Rosenberg, N. A., Pritchard, J. K., Weber, J. L., et al. (2002). Genetic structure of human populations. *Science*, **298**, 2381–85.

Rowley, J. D. (1973). A new consistent chromosomal abnormality in chronic myelogenous leukaemia identified by quinacrine fluorescence and Giemsa staining. *Nature*, **243**, 290–93.

Ruse, M. (2009). The history of evolutionary thought. In M. Ruse and J. Travis (eds.), *Evolution: the first four billion years*, 1–48. Belknap Press of Harvard University Press, Cambridge, MA.

Ryan, F. (2001). *Darwin's blind spot: evolution beyond natural selection*. Houghton Mifflin, Boston.

Sacks, F. M., Svetkey, L. P., Vollmer, W. M., et al. (2001). Effects on blood pressure of reduced dietary sodium and the Dietary Approaches to Stop Hypertension (DASH) diet. DASH-Sodium Collaborative Research Group. *New England Journal of Medicine*, **344**, 3–10.

Saleheen, D. and Frossard, P. M. (2008). The cradle of the deltaF508 mutation. *Journal of Ayub Medical College, Abbottabad*, **20**, 157–60.

Sapp, J. (1994). *Evolution by association: a history of symbiosis*. Oxford University Press, New York.

Scally, A. and Durbin, R. (2012). Revising the human mutation rate: implications for understanding human evolution. *Nature Reviews Genetics*, **13**, 745–53.

Scherf, A., Lopez-Rubio, J. J., and Riviere, L. (2008). Antigenic variation in *Plasmodium falciparum*. *Annual Review of Microbiology*, **62**, 445–70.

Schirru, E., Corona, V., Usai-Satta, P., et al. (2007). Decline of lactase activity and c/t-13910 variant in Sardinian childhood. *Journal of Pediatric Gastroenterology and Nutrition*, **45**, 503–6.

Schrag, S. J. and Wiener, P. (1995). Emerging infectious disease: what are the relative roles of ecology and evolution? *Trends in Ecology and Evolution*, **10**, 319–24.

Schug, T. T., Janesick, A., Blumberg, B., et al. (2011). Endocrine disrupting chemicals and disease susceptibility. *Journal of Steroid Biochemistry and Molecular Biology*, **127**, 204–15.

Schweber, S. S. (1980). Darwin and the political economists: divergence of character. *Journal of the History of Biology*, **13**, 195–289.

Scriver, C. R. (1984). The Canadian Rutherford lecture. An evolutionary view of disease in man. *Proceedings of the Royal Society B: Biological Sciences*, **220**, 273–98.

Scriver, C. R. and Waters, P. J. (1999). Monogenic traits are not simple: lessons from phenylketonuria. *Trends in Genetics*, **15**, 267–72.

Shammas, M. A. (2011). Telomeres, lifestyle, cancer, and aging. *Current Opinion in Clinical Nutrition and Metabolic Care*, **14**, 28–34.

Sharp, P. M. and Hahn, B. H. (2011). Origins of HIV and the AIDS pandemic. *Cold Spring Harbor Perspectives in Medicine*, **1**, a006841.

Silverman, F. N. (1993). À la recherche du temps perdu and the thymus (with apologies to Marcel Proust). *Radiology*, **186**, 310–11.

Simoons, F. J. (1978). The geographic hypothesis and lactose malabsorption. A weighing of the evidence. *American Journal of Digestive Diseases*, **23**, 963–80.

Sindic, A. and Schlatter, E. (2006). Cellular effects of guanylin and uroguanylin. *Journal of the American Society of Nephrology*, **17**, 607–16.

Sloboda, D. M., Beedle, A. S., Cupido, C. L., et al. (2009). Impaired perinatal growth and longevity: a life history perspective. *Current Gerontology and Geriatric Research*, article 608740.

Smith, D. L., McKenzie, F. E., Snow, R. W., et al. (2007). Revisiting the basic reproductive number for malaria and its implications for malaria control. *PLoS Biology*, **5**, e42.

Sontag, S. (1978). *Illness as metaphor*. Farrar, Straus and Giroux, New York.

Spencer, H. (1864). *Principles of biology*. Williams and Norgate, London.

Stamatoyannopoulos, J. A. (2012). What does our genome encode? *Genome Research*, **22**, 1602–11.

Stearns, S. C. (1992). *The evolution of life histories*. Oxford University Press, Oxford.

Stearns, S. C. (2012). Evolutionary medicine: its scope, interest and potential. *Proceedings of the Royal Society B: Biological Sciences*, **279**, 4305–21.

Stearns, S. C., Allal, N., and Mace, R. (2008). Life history theory and human development. In C. Crawford and D. Krebs (eds.), *Foundations of evolutionary psychology*, 47–69. Lawrence Erlbaum Associates, New York.

Stearns, S. C., Byars, S. G., Govindaraju, D. R., et al. (2010). Measuring selection in contemporary human populations. *Nature Reviews Genetics*, **11**, 611–22.

Stearns, S. C. and Koella, J. C., eds. (2008). *Evolution in health and disease* 2nd edn. Oxford University Press, Oxford.

Sterelny, K. and Griffiths, P. E. (1999). *Sex and death: an introduction to philosophy of biology*. University of Chicago Press, Chicago.

Stewart, S. T., Cutler, D. M., and Rosen, A. B. (2009). Forecasting the effects of obesity and smoking on U.S. life expectancy. *New England Journal of Medicine*, **361**, 2252–60.

Stiehm, E. R. (2006). Disease versus disease: how one disease may ameliorate another. *Pediatrics*, **117**, 184–91.

Stoll, N. R. (1999). This wormy world. 1947. *Journal of Parasitology*, **85**, 392–96. Originally published in *Journal of Parasitology*, **33**, 1–18, 1947.

Strachan, D. P. (1989). Hay fever, hygiene, and household size. *BMJ*, **299**, 1259–60.

Suryadinata, R., Sadowski, M., and Sarcevic, B. (2010). Control of cell cycle progression by phosphorylation of cyclin-dependent kinase (CDK) substrates. *Bioscience Reports*, **30**, 243–55.

Talman, A. M., Domarle, O., McKenzie, F. E., et al. (2004). Gametocytogenesis: the puberty of *Plasmodium falciparum*. *Malaria Journal*, **3**, 24.

Tattersall, I. (2009). Human origins: out of Africa. *Proceedings of the National Academy of Sciences, USA*, **106**, 16 018–21.

Thurlow, L. R., Joshi, G. S., and Richardson, A. R. (2012). Virulence strategies of the dominant USA300 lineage of community-associated methicillin-resistant *Staphylococcus aureus* (CA-MRSA). *FEMS Immunology and Medical Microbiology*, **65**, 5–22.

Tinbergen, N. (1963). On the aims and methods of ethology. *Zeitschrift fur Tierpsychologie*, **20**, 410–33.

Tishkoff, S. A., Reed, F. A., Ranciaro, A., et al. (2007). Convergent adaptation of human lactase persistence in Africa and Europe. *Nature Genetics*, **39**, 31–40.

Trevathan, W. R., Smith, E. O., and McKenna, J. J., eds. (2008). *Evolutionary medicine and health: new perspectives*. Oxford University Press, Oxford.

Tsai, H. C. and Baylin, S. B. (2011). Cancer epigenetics: linking basic biology to clinical medicine. *Cell Research*, **21**, 502–17.

Turnbaugh, P. J., Ley, R. E., Hamady, M., et al. (2007). The human microbiome project. *Nature*, **449**, 804–10.

References 157

Varmus, H. E. (1990). Nobel lecture. Retroviruses and oncogenes. I. *Bioscience Reports*, **10**, 413–30.

Vesa, T. H., Marteau, P., and Korpela, R. (2000). Lactose intolerance. *Journal of the American College of Nutrition*, **19**, 165S–75S.

Vineis, P. and Berwick, M. (2006). The population dynamics of cancer: a Darwinian perspective. *International Journal of Epidemiology*, **35**, 1151–59.

Vogelstein, B. and Kinzler, K. W. (1993). The multistep nature of cancer. *Trends in Genetics*, **9**, 138–41.

Wallace, D. C. (2010). Mitochondrial DNA mutations in disease and aging. *Environmental and Molecular Mutagenesis*, **51**, 440–50.

Walsh, S. J. and Rau, L. M. (2000). Autoimmune diseases: a leading cause of death among young and middle-aged women in the United States. *American Journal of Public Health*, **90**, 1463–66.

Walter, J. and Ley, R. (2011). The human gut microbiome: ecology and recent evolutionary changes. *Annual Review of Microbiology*, **65**, 411–29.

Watt, G. C. (1996). All together now: why social deprivation matters to everyone. *BMJ*, **312**, 1026–29.

Weinstock, J. V. and Elliott, D. E. (2009). Helminths and the IBD hygiene hypothesis. *Inflammatory Bowel Diseases*, **15**, 128–33.

Weiss, F. U., Simon, P., Bogdanova, N., et al. (2005). Complete cystic fibrosis transmembrane conductance regulator gene sequencing in patients with idiopathic chronic pancreatitis and controls. *Gut*, **54**, 1456–60.

Weiss, K. M. and Buchanan, A. V. (2009). The cooperative genome: organisms as social contracts. *International Journal of Developmental Biology*, **53**, 753–63.

Weiss, R. A. (2002). Virulence and pathogenesis. *Trends in Microbiology*, **10**, 314–17.

Wells, J. C. (2010). Maternal capital and the metabolic ghetto: an evolutionary perspective on the transgenerational basis of health inequalities. *American Journal of Human Biology*, **22**, 1–17.

Wells, J. C. (2012). Obesity as malnutrition: the role of capitalism in the obesity global epidemic. *American Journal of Human Biology*, **24**, 261–76.

Wells, J. C. and Siervo, M. (2011). Obesity and energy balance: is the tail wagging the dog? *European Journal of Clinical Nutrition*, **65**, 1173–89.

Welsh, M. J., Ramsey, B. W., Accurso, F., et al. (2001). Cystic fibrosis. In D. Valle (ed.), *Scriver's OMMBID: the online metabolic & molecular bases of inherited disease*. McGraw-Hill, New York. <http://www.ommbid.com/>, accessed 30 September 2012.

Westendorp, R. G. and Kirkwood, T. B. (1998). Human longevity at the cost of reproductive success. *Nature*, **396**, 743–46.

Wilkinson, R. G. (2001). *Mind the gap*. Yale University Press, New Haven, CT.

Williams, G. C. (1957). Pleiotropy, natural selection, and the evolution of senescence. *Evolution*, **11**, 398–411.

Williams, G. C. and Nesse, R. M. (1991). The dawn of Darwinian medicine. *Quarterly Review of Biology*, **66**, 1–22.

Wiuf, C. (2001). Do delta F508 heterozygotes have a selective advantage? *Genetical Research*, **78**, 41–47.

World Health Organization (2006). Constitution of the World Health Organization, <http://www.who.int/governance/eb/who_constitution_en.pdf>, accessed 30 September 2012.

Wrangham, R. (2009). *Catching fire: how cooking made us human*. Basic Books, New York.

Young, J. H. (2007). Evolution of blood pressure regulation in humans. *Current Hypertension Reports*, **9**, 13–18.

Zampieri, F. (2009). Medicine, evolution, and natural selection: an historical overview. *Quarterly Review of Biology*, **84**, 333–55.

Zhu, W. X., Lu, L., and Hesketh, T. (2009). China's excess males, sex selective abortion, and one child policy: analysis of data from 2005 national intercensus survey. *BMJ*, **338**, b1211.

zur Hausen, H. (2002). Papillomaviruses and cancer: from basic studies to clinical application. *Nature Reviews Cancer*, **2**, 342–50.

Index

accidents 14–15, 53
adaptations 5–7, 9, 41, 55, 90
aging 11, 19–21, 51–63, 128, 139–40
    antagonistic pleiotropy 57, 62–3
    balance, loss of 60
    disposable soma hypothesis 57
agricultural revolution 22–5
Allison, Anthony 2
among-host selection 83
Andersen, Dorothy 44
Anopheles gambiae 111–12
antibiotic resistance 2, 98
    Staphylococcus aureus 87–9
appendix 1
Aristotle 55
asthma 134–7
autoimmune diseases 134–7
    multiple sclerosis 134–6
    rheumatic fever 87, 134

Barker, David 63
basic reproductive number (*Ro*) 78–9, 83, 89
    of *Plasmodium falciparum* 108–9
    of sexually transmitted pathogens 92–3
Bernard, Claude 9
between-host selection, *see* among-host selection
birth rates 13–16, 22–3, 26
birth weight 15, 41, 62–3, 140
bottlenecks, population 32, 129
breastfeeding 24, 52, 120
burden of disease 26, 43, 54, 98–9, 128–31
    disability-adjusted life-years lost 81

cancer 14, 26, 53, 65–76
    cancer stem cells 72
    carcinogenesis 74–6
    colon cancer 71, 74
    evolution and ecology of 71–4
    hallmarks of cancer 71–2
    oncogenes 70, 75
    tumor suppressor genes 70, 72, 74
cardiovascular diseases 14, 26, 53, 63; *see also*
    myocardial infarction; stroke
cataracts 58–9
causes of death 14–15, 21–3, 26, 53–4, 134; *see also*
    specific diseases
    extrinsic (external) 55–7
cell cycle 66–70

chicken pox 83, 87
cholera 45, 90
Clostridium tetani 85–6
Cohen, Joel:
    How Many People Can the Earth
        Support? 21–3
color vision 34–5
competitive exclusion 81, 88
    in cancer 76
crowd diseases 78, 94
cystic fibrosis 43–9
    cystic fibrosis transmembrane conductance
        regulator (CFTR) 44–9, 90

dairying, *see* milk, milking
Darwin, Charles 1, 3–8
    On the Origin of Species 3
    The Descent of Man 1
    The Variation of Animals and Plants under
        Domestication 1
dead-end evolution 86, 88, 93, 95, 97
death rates, *see* mortality rates
degenerative diseases 26, 53–4, 128
demographic transition 23–6, 54
developmental abnormalities 15, 53
developmental origins of health and disease 62–3, 140
diabetes 53, 63, 128–31
    insulin-dependent 134
    thrifty genotype hypothesis 128–9
diet 128–31
    Paleolithic diet 129
diseases of affluence, *see* man-made diseases
disgust response 81, 98, 137
DNA repair 31, 58, 60
    and cancer 66, 68, 70, 72, 74–6
Duffy antigen 113, 137

ENCODE project 29
endocrine disruptors 63
environment of evolutionary adaptedness (EEA) 127
epidemiologic transition 26, 54, 128
epigenetic regulation 37–8
    and development 62–3
    in cancer 66, 69–72
    of lactase expression 120
epistasis 35–6
    in cystic fibrosis 47, 49
evidence-based medicine 8, 130

fertility rates:
  age-specific 13, 16, 56
  replacement rate 18–19, 22, 24
  total fertility rate (TFR) 16, 18–19, 22, 24, 26
fever 90
  in malaria 106
Fisher, R. A. 33, 36–7
fitness 5–12; *see also* inclusive fitness
Flexner Report 3
fomites 80, 88–9
founder effect 32, 39
  in cystic fibrosis 49
frequency dependent selection 36–7, 81

Garrod, Sir Archibald 43
gene–culture coevolution 115–25
gene flow 11, 30, 32, 39–40
genetic dominance 33–4
genetic drift 11, 30–2, 39–41, 128
genetic linkage, *see* hitchhiking, genetic
genomic imprinting 38
glucose-6-phosphate dehydrogenase
    deficiency 2, 113

Haldane, J. B. S. 2, 10, 81
Hamilton, William 5, 55–6
haplotype analysis 36, 41
  CFTR 48–9
  hemoglobin S 113
  lactase persistence 123
Hardy-Weinberg model 38–9
hay fever 135
health disparities 137–41
helminths 22, 77, 135–7
hemochromatosis 36
hemoglobin S, *see* sickle cell hemoglobin
herd immunity 79
heterozygote advantage 34–6, 112
high-fructose corn syrup 130
hitchhiking, genetic 36, 49
HIV/AIDS 14, 92–3, 98–102, 142
host-pathogen coevolution 7, 12, 77–90
Human Genome Project 10, 30
human papilloma virus (HPV) 75–6
hygiene hypothesis 134–7
hypertension 12, 62, 131–4

inclusive fitness 5, 93
industrial revolution 23, 40
infanticide 22, 27
infant mortality 14–15, 18–20, 22–3
infectivity, *see* transmissibility
inflammatory bowel disease 136
influenza 87

Jackson, John Hughlings 1

kidney disease 60, 62

lactase 115–25
  persistence 120–5
  restriction 120–5
lactose 115–25
  intolerance 117
life cycle 7, 51–4
life expectancy 18, 20, 26, 61
life history theory 11, 51–63
  strategies 24, 59, 62, 94, 140
  tradeoffs 24, 52, 55–61, 63
Lyme disease 86

McKeown, Thomas:
  The Origins of Human Disease 21–3, 26
macroevolution 7, 12
major histocompatibility complex
    (MHC) 36
  polymorphisms 81, 113–14
malaria 2, 14, 103–14
malaria hypothesis 2
malnutrition 12, 14–15
Malthus, Thomas 4, 23, 68
man-made diseases 26, 54, 127–42
  allergic and autoimmune diseases 134–7
  diabetes 128–31
  health disparities 137–41
  hypertension 131–4
Marx, Karl 8
Mayr, Ernst 3
  biological species concept 4, 32
measles 79, 83, 87
Medawar, Peter 57
Mendel, Gregor 29, 35, 43
microbiome 9–10, 38, 77, 81
  intestinal 119, 122, 130, 135–6
  vaginal 94
microevolution 7, 12
migration, *see* gene flow
milk, milking 116–19
mismatch:
  between ancestral and current environments 12,
    25, 128–31
  between fetal and postnatal environments 63
mitochondria 59
modern synthesis 3, 29
modifier genes, *see* epistasis
Monod, Jacques 6
mortality rates 13–15, 22, 26
  age-specific 19–20, 56
  infant mortality 14–15, 18–20, 22–3

multiple sclerosis 134–6
mumps 83
mutagens:
    and cancer 75–6
mutation 11
    C→T transitions 31, 37, 46, 70, 72, 122
    somatic 61
mutation rate, human 31, 34
    somatic 69
mutation-selection balance 33–4, 43, 48, 56
Mycobacterium tuberculosis, see tuberculosis
myocardial infarction 14, 131, 141
myxomatosis 84–5

Neisseria gonorrhoeae 95
nematodes 135–7
Nesse, Randolph 2
niche construction 116, 123
norm of reaction 52, 62
nursing, see breastfeeding

obesity 12, 128–31
oncogenes 70, 75
Online Mendelian Inheritance in Man (OMIM) 43
ontogeny 9; see also developmental origins of health
    and disease
otoconia 60–1

p53 67–70, 76
Paleolithic era 129
parental investment 24–5
pathogens:
    basic reproductive number (Ro) 78–9, 83, 89
    biology 77, 80
    fitness 82–4, 86
    life histories 82–4
    virulence 80–1
phylogeny 9
physiological capital 59–63, 140
Plasmodium falciparum 103–14
    evolution 110–11
    life cycle 103–5, 107–8, 110
Plasmodium vivax 103, 113
plasticity, developmental 61–3
pleiotropy 35–6, 44, 52, 83, 137; see also aging
Pleistocene epoch 115, 127
poliovirus 86–7
polymorphism, polymorphic genes 30, 33–4, 36–7,
    136–7
population growth rates 13–6, 19
    human population growth 21–3, 26, 121
population health 141–2
population pyramids 16–17, 55
prematurity 15, 62, 140

proximate cause 9–12, 14
    of aging 58–9
    of cystic fibrosis 48
    of health disparities 140
puberty 51

Ro, see basic reproductive number
rabies virus 86
renal disease, see kidney disease
replacement rate 78–9
retinoblastoma protein (pRb) 67, 70,
    74, 76
rheumatic fever 87, 134
rhinoviruses 82, 84

salt 131–4
sex ratio 18, 27
    of Plasmodium falciparum 105
sexually transmitted diseases 91–102
sexual selection 6, 18
short-sighted evolution, see dead-end evolution
sickle cell anemia 2, 112–13
sickle cell hemoglobin 112–13
signals of selection 41
skin pigmentation 40–1
smallpox 79, 95
smoking, tobacco 62, 75
socioeconomic disparities 15, 25, 127; see also
    health disparities
sodium, see salt
somatic repair 51, 55, 58–61, 140; see also
    DNA repair
Spencer, Herbert 5
stabilizing selection 41, 101
Staphylococcus aureus:
    methicillin resistance 87–9
stem cells 60, 68–9, 72, 120
stroke 14, 128, 131, 134, 141
struggle for existence 4
sucrose 130
suicide 14, 53
survival of the fittest 5
survival rates, age-specific 13
survivorship curves 20–1
sweat 44–5, 131
syphilis 95–8

telomeres, telomerase 59, 69
thalassemia 2, 35–6, 112–13
Tinbergen, Nikolaas 9
total fertility rate (TFR) 16, 78–9
tradeoffs 8–9, 12, 24, 35
    in host responses to pathogens 90
    in life history traits 24, 52, 55–61, 63

transforming growth factor β1 (TGFβ1)
  47, 68
transmissibility 80–1
Treponema pallidum, see syphilis
tuberculosis 14, 48–9, 83
tumor suppressor genes 70, 72, 74

ultimate cause 9–12, 14
  of cystic fibrosis 48
urbanization 22–5

Vibrio cholerae, see cholera
violence 14, 53
  fear of 140
Virchow, Rudolf 47

virulence 12, 80–1
  of *Plasmodium falciparum* 109–10
vitamin D 40, 123, 137

war 14
weaning 52–3, 118, 120–1, 123
Western diseases, *see* man-made diseases
Williams, George 2, 57
within-host selection 82–3, 93
  of HIV 101
  of *Plasmodium falciparum* 109
World Health Organization (WHO) 8, 13–14
worms, *see* helminths

zoonoses 23